To Géraldine, in memory of Elad

To my grandfather Lala

To his raw relationship with nature

his humor and madness

his hatred of injustice

his unwillingness to compromise

MY HEARTFELT THANKS

to Lily and Philippe Staib, without whose generosity

this book never would have come to light.

Thanks to Pierre Naim for having offered me

the precious freedom to realize my projects.

CLAIRE NOUVIAN

THE UNIVERSITY OF CHICAGO PRESS

CHICAGO AND LONDON

THE DEEP

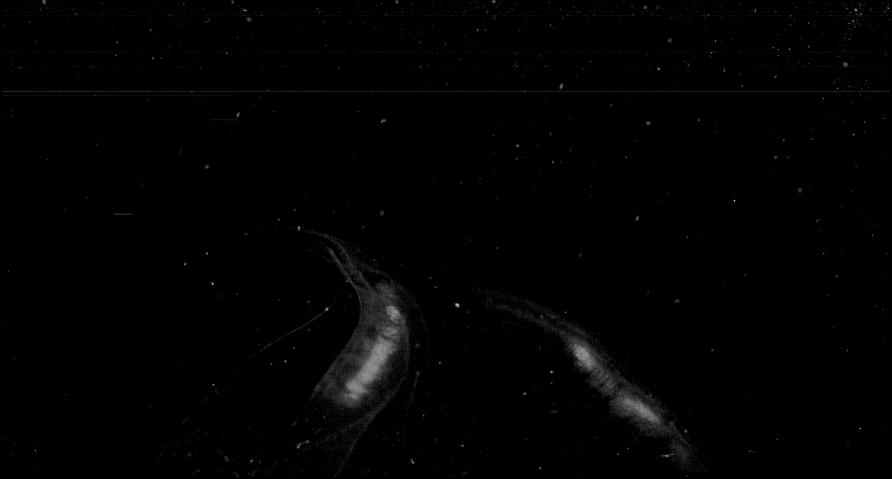

LIFE IN THE WATER COLUMN

CONTENTS

The University of Chicago Press, Chicago 60637
The University of Chicago Press, Ltd., London

© 2007 by Éditions Fayard

All rights reserved. Published 2007

Printed in France

5 th printing

16 15 14 13 12 11 10 09 08 07   1 2 3 4 5
ISBN-13: 978-0-226-59566-5 (cloth)
ISBN-10: 0-226-59566-8 (cloth)
CIP data on file

LIFE AT THE BOTTOM

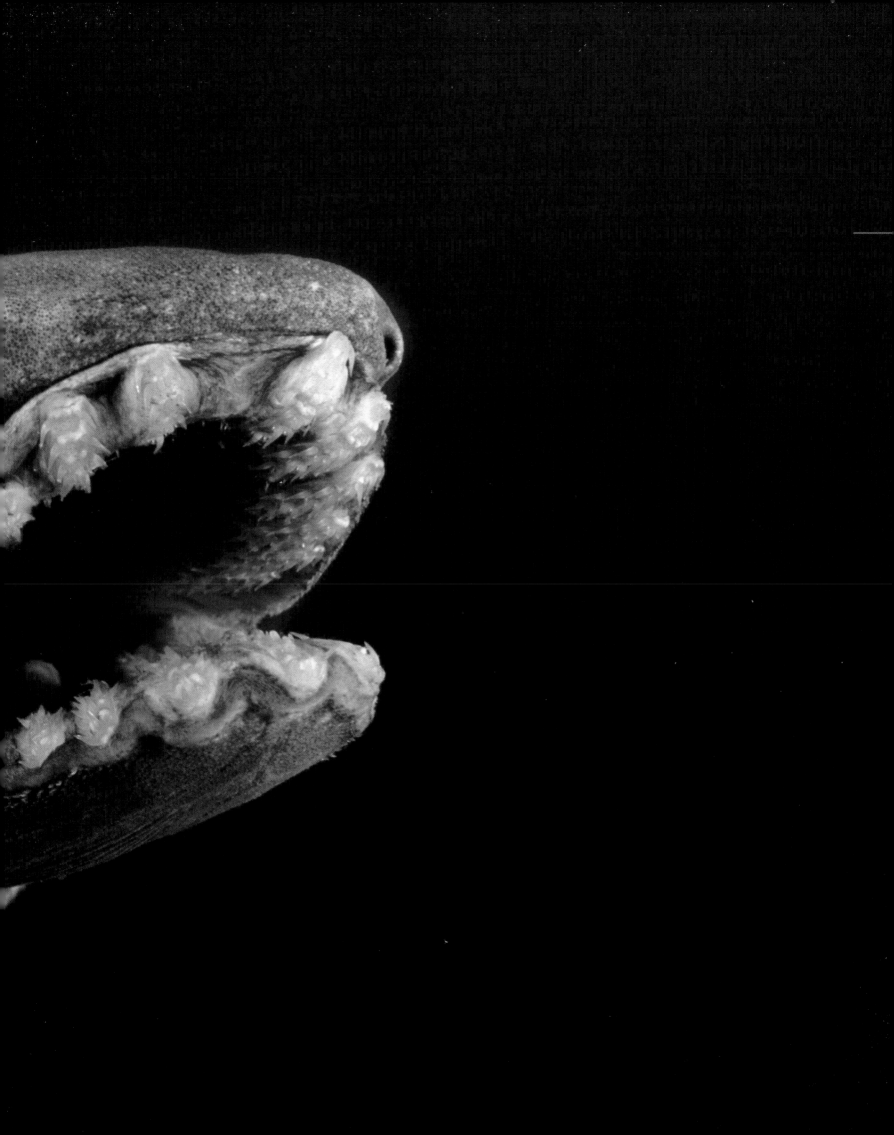

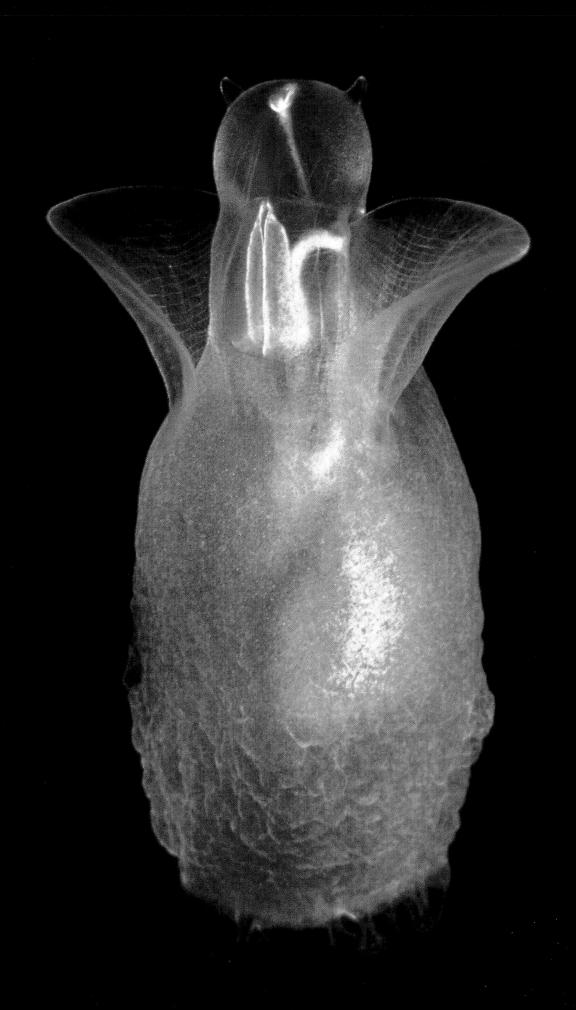

In 2001, after watching a stunningly beautiful film at the Monterey Bay Aquarium in California, I took my first dive into the deep. In that one moment, and without expecting it in the least, my life changed direction.

The images showed animals that had been filmed in the depths of the Monterey Canyon by the Monterey Bay Aquarium Research Institute. Fabulous creatures had been discovered there, some with surprising shapes or baffling colors, others that spat out threatening flashes of blue light, and others still that undulated with infinite grace, producing iridescent sparkles.

preface

Strange animals without head or tail twisted and unwound, like fluid ribbons in some sensuous sort of dance. One struck me more than the others: a dark red, transformist octopus, sporting two large, comical "ears" that made it look all the more endearing. It seemed almost to float within space, among the stars... With majestic elegance and exaggerated slowness, it inflated its mantle into a balloon, or flattened it out into the shape of a disk. Other times, it lifted its skin over its head, to metamorphose into an abyssal pumpkin, then reappearing suddenly, this time twisted lengthwise, looking like a morning glory, folded up, nice and tidy, for the night. Then, at last, it drifted away, tranquilly, into the darkness of its own universe...

What a universe! I watched the film four times in a row, scrutinizing the screen incredulously, searching for some clue that inevitably would betray even the best special effects artist. My thoughts looped, just like the film: "This isn't possible... These animals aren't real... These effects are magnificently executed...totally unheard of; Tim Burton must be behind all this!" And as my eye gradually confirmed that what I was seeing was not the mark of computer-generated images, but rather the inimitable texture of reality, my thoughts began to change: "How is it possible that the Earth bears such marvels and that people don't even know about them? Why doesn't somebody stop the world, for a minute, or maybe just a *second*, to announce that these creatures exist down here, at the very bosom

OPPOSITE
Cliopsis krohni
Sea angel
_SIZE 4 cm
DEPTH 1500 m

In spite of its innocent appearance, there is nothing angelic about this swimming snail; it is an active predator that devours other midwater snails using a tongue studded with small, sharp teeth. Curiously, its mouth is at the top of its head.

The vampire squid, in spite of its name, is actually a harmless animal that lives suspended in the dark strata of the oceans, sometimes unfurled in this typical umbrella posture. Like other animals living at great depths, its musculature is much reduced, though it is capable of surprising bursts of speed over short distances. When fleeing an enemy, it may create confusion by activating its various bioluminescent organs.

of our planet?" I felt the same emotion, the same shock, as though I had just learned that the first extraterrestrials had been discovered ... and filmed!

I was dazzled... speechless ... astounded...

As crazy as it might seem, I had fallen in love at first sight. Like an adolescent surprised by the power of love, my life had suddenly taken on new dimensions. It was as though a veil had been lifted, revealing unexpected points of view, vaster and more promising. On the plane back to Paris, all my thoughts were focused on one subject: the deep sea. I imagined this colossal volume of water, cloaked in permanent darkness, and I pictured the fantastic creatures that swam there, far from our gazes, the surrealist results of an ever inventive Nature.

Soon, the images started fading away, quickly and quietly, and I was left with painful disappointment when I realized how little documentation there was on the topic. The animals in the film were magnificent; they were *life*, but the fleeting nature of those animated images did not satisfy my curiosity.

First of all, I needed a book that would allow me, at my own pace and for hours on end if I wished, to examine these incredible beings that lived secluded in the oceans' darkness. I needed a bit of distance, and some liberty, to start on my own inner journey and travel mentally with these aliens from inner space,

throbbing and pulsating at the heart of a liquid sky. I wanted to know everything about them: how deep they lived, how they reproduced, their dimensions, their names... I dreamed of a book that would bring together all the most beautiful images captured in that deep-sea frontier, inaccessible to the majority of us; a book that would give voice to those who held the precious information about this singular environment, a book that would be accessible to all. In short, I dreamed of a book that would bring the deep to light!

I wanted this book so earnestly, and I had such a precise idea of the mission it ought to fulfill, that I decided to realize it myself. I called upon researchers from all over the world, the big names of current oceanography, and one after the other, they each agreed spontaneously to collaborate on my project.

This book corresponds exactly to my vision; it "feeds" both aesthetic and intellectual senses. Moreover, it is a work that, I hope, will have the strength to captivate us and to remind us that we belong to a vast, living chain that is incredibly beautiful, terribly fragile, and—until proved otherwise—miraculously unique in

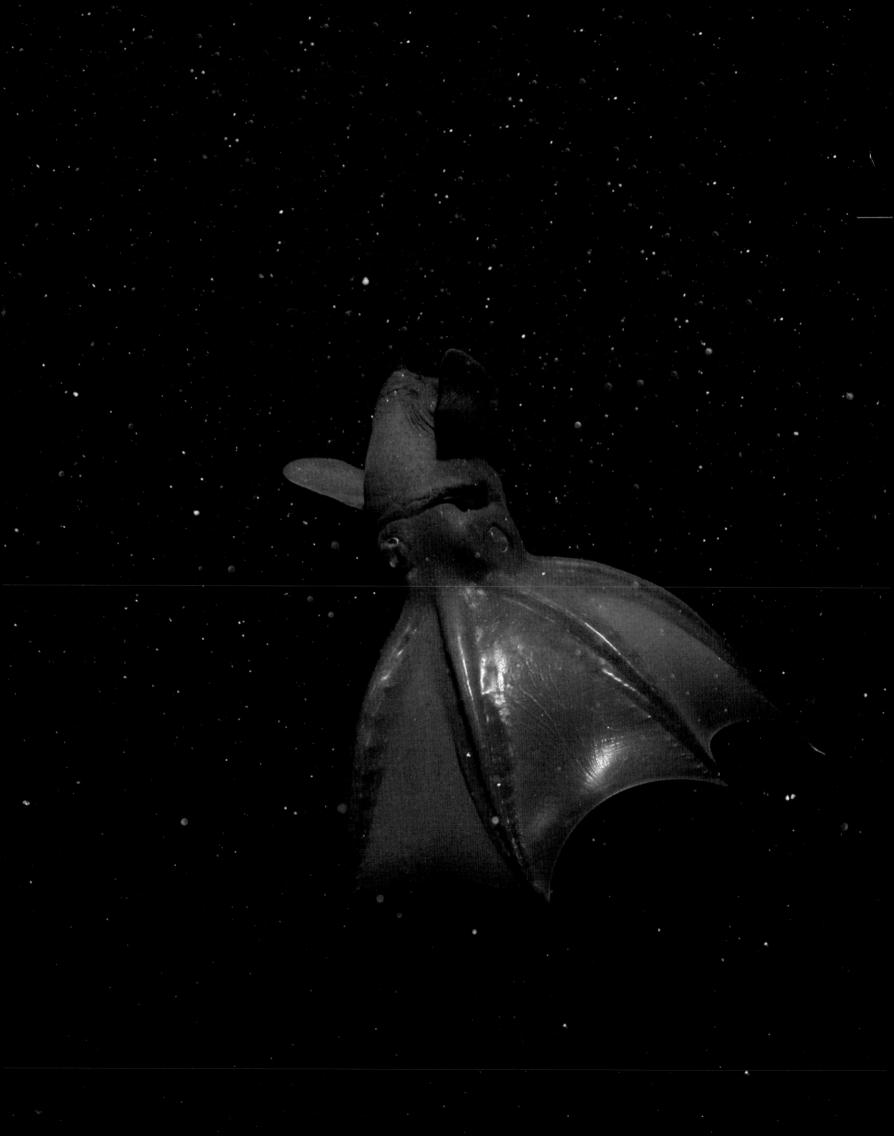

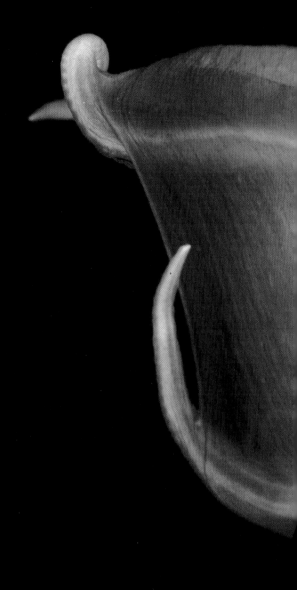

RIGHT

Stauroteuthis syrtensis

Glowing sucker octopus

_SIZE up to 50 cm
DEPTH down to 2500 m

The deep-sea cirrate octopuses are nicknamed the "Dumbo octopuses" because of their two fins, which resemble elephant ears. They have two main modes of locomotion: they can beat their fins or propel themselves through the water by rhythmically contracting their mantle. To date, the glowing sucker octopus has been observed only in the Atlantic Ocean.

NEXT DOUBLE PAGE SPREAD

Mertensia ovum

_SIZE 8 cm

Comb jellies, or ctenophores, differ from true jellies in their method of snaring the small creatures they feed upon. Rather than capture prey using poison-tipped harpoons, they deploy sticky cells that literally glue the comb jelly's tentacles to small crustaceans, larvae, and other planktonic animals. Its two orange tentacles extend and retract to pass food to its mouth.

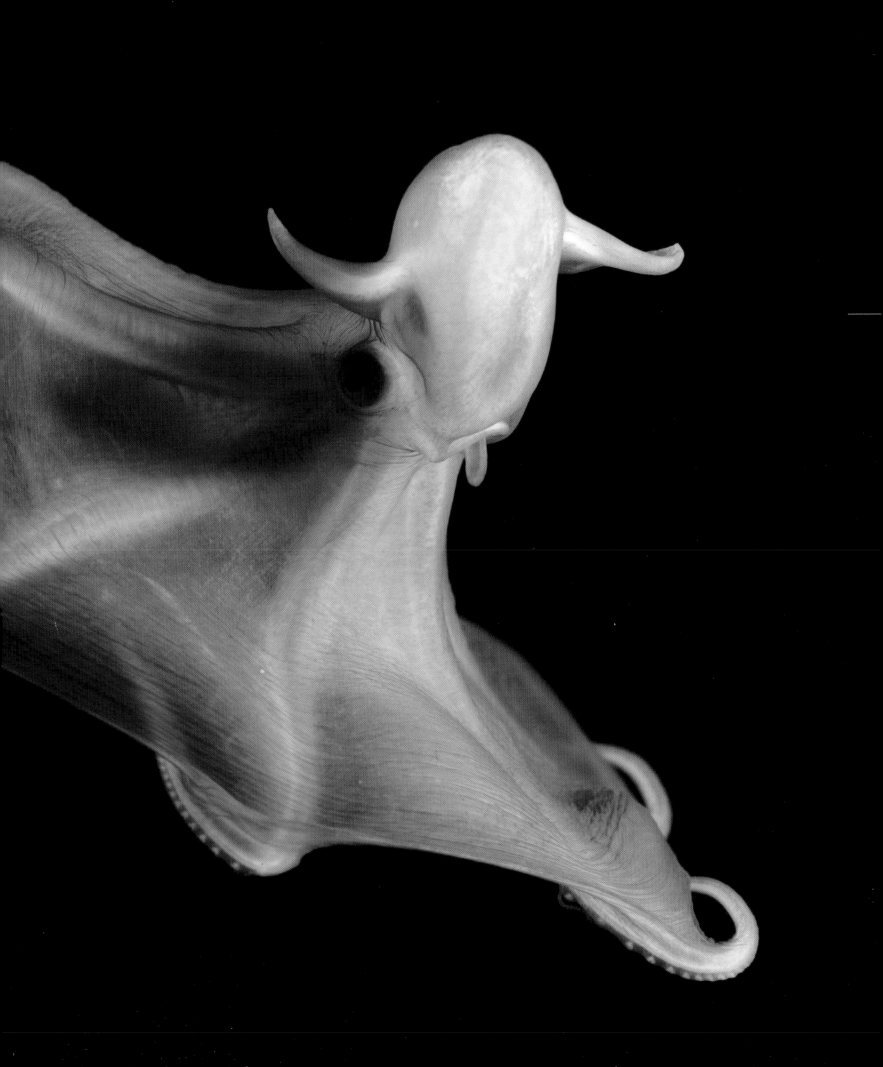

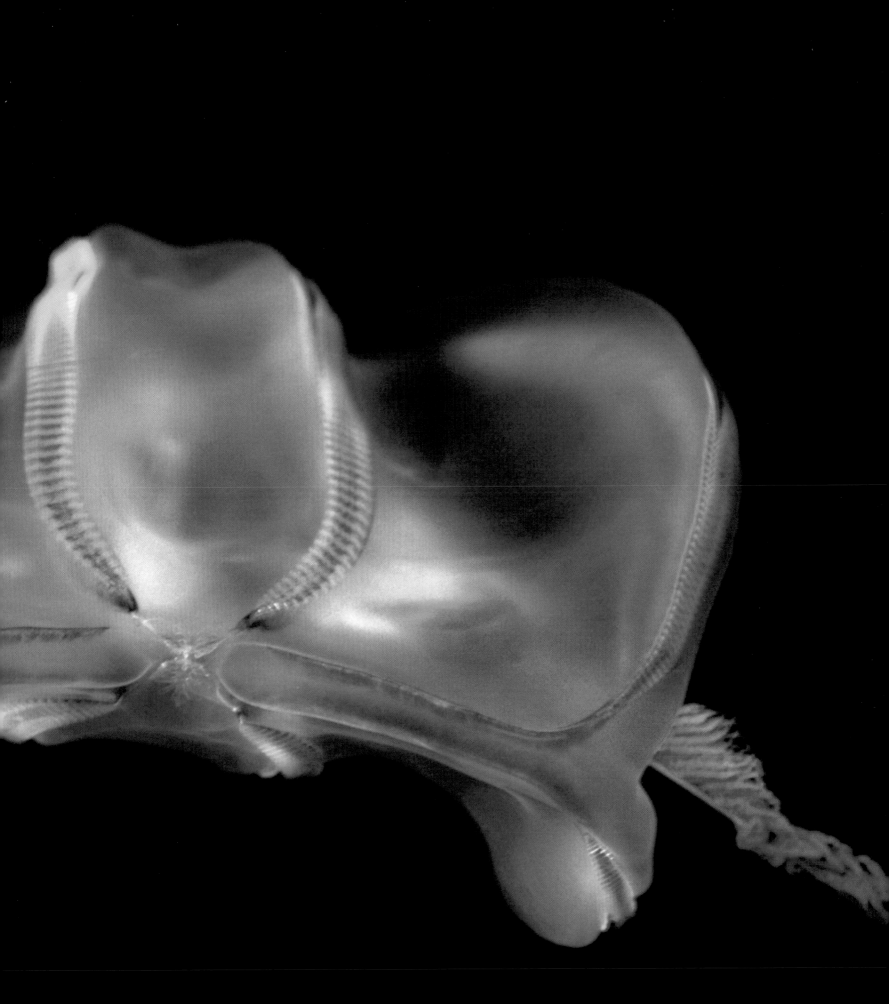

> "On a planetary scale, birds crawl."
>
> Jacques-Yves Cousteau

On dry land, most organisms are confined to the surface, or at most to altitudes of a hundred meters—the height of the tallest trees. In the oceans, though, available living space has both vertical and horizontal dimensions. With an average depth of 3800 meters, the oceans offer 99% of the space where life can develop on Earth. A staggering thought.

The deep sea, which has been immersed in total darkness since the dawn of time, occupies 85% of this space, and thus forms the planet's largest habitat.

And what do we know about it?

Compared to what remains to be discovered, practically nothing. The earliest explorations of the midocean ridges date back to the 1970s. The first midwater dives to explore the vast deep-sea

EaRTH's LaRgEST LiVing REaLm

domain took place in the 1980s; even the very first studies of the deep seafloor were undertaken only relatively recently, with the large-scale oceanographic trawling campaigns of the nineteenth century. Currently, only about 5% of the seafloor has been mapped with any reasonable degree of detail, which relegates the overwhelming majority of the abyssal plains and other deep-sea habitats to the unknown. Moreover, in some of the expeditions carried out in the Southern Atlantic or around seamounts in the Pacific, 50% to 90% of the specimens coming up in the nets are unidentified specimens. For the last twenty-five years, a new deep-sea species has been described every two weeks, on average. Current estimates about the number of species yet to be discovered vary between 10 and 30 million. By comparison, the number of known species populating the planet today, whether terrestrial, aerial, or marine, is estimated at about 1.4 million. The deep sea no longer has anything to prove; it is without doubt Earth's largest reservoir of life.

Contemporaries of Jules Verne, keen on classifications of all sorts, believed they had almost completed the exploration of the planet and that they could even fit the descriptions of their discoveries into a few volumes of a single encyclopedia. The study of the deepwater world has radically altered this point of view. The first circumnavigation of the globe—the nineteenth-century *Challenger* expedition—alone required fifty volumes to document the results

OPPOSITE
Galiteuthis phyllura
Cockatoo squid
_SIZE up to 2.7 m
DEPTH 300–1400 m

The only part of the cockatoo squid's body that isn't transparent is its eyes. To camouflage its silhouette from predators below, the squid uses its bioluminescent photophores, which pivot so that they are always directed downward; these disguise the opacity of its eyes by counterillumination. While at rest, the cockatoo squid holds its arms and tentacles in a sort of tuft over its head, and so resembles the bird whose name it bears. This young specimen is about 15 cm long, but as an adult it may reach nearly 3 m.

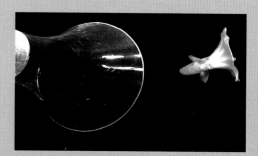

ABOVE
Cirroteuthis muelleri
Dumbo octopus
_SIZE 30 cm
DEPTH 4000 m

There is nothing like a flared suction sampler for harvesting the fragile creatures populating the midwater. The sampler works like a deep-sea vacuum cleaner, enabling researchers to collect animals and bring them to the surface in near-perfect condition, significantly facilitating their work of identification.

> "At a time when most think of outer space as the final frontier, we must remember that a great deal of unfinished business remains here on Earth." Robert D. Ballard

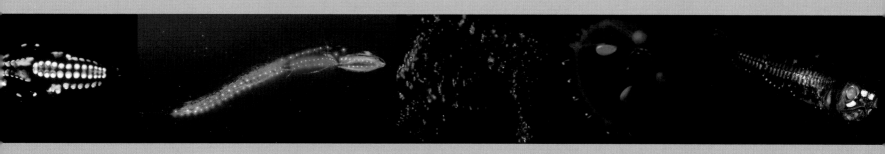

OPPOSITE

Scrippsia pacifica

Giant bell jelly

_SIZE 10 cm in height
DEPTH 400 m

The dominance of gelatinous organisms in the deepwater ecosystem was unsuspected before deep-sea submersibles permitted researchers to explore the water column. Before that, whenever an occasional specimen of the *Scrippsia* jelly would reach the surface, having strayed from its kingdom in the depths, it arrived in such poor condition that its natural beauty was brutally marred.

ABOVE

Bioluminescence is based on a chemical reaction producing a "cold" light that consumes very little energy. Deep-sea species employ their own lights for a wide variety of purposes: the deep-sea silver hatchetfish (*Argyropelecus olfersi*, FIRST PHOTO FROM LEFT) and the bristlemouth (*Cyclothone* SECOND FROM LEFT) use glow for camouflaging counterillumination, while the headlight lanternfish (*Diaphus* sp. FAR RIGHT) sends out beams of light to detect prey. One of the functions of this light common to all is its defensive role.

obtained on the four-year journey. This expedition marked the point of departure for all subsequent oceanographic missions; to date, each brings back marvelous treasures from the deep, each pushes the frontiers of knowledge a bit farther.

Without any doubt, we still live in an age of exploration, possibly the greatest of all time. The difference between today's adventures and those undertaken by Columbus or Livingstone lies in the equipment: submersibles and remote-controlled robots have replaced caravels and slide rules. The biggest surprise of our era has been that the most startling discoveries of the twentieth century were not waiting for us in space, as we would have expected, but at the very heart of our oceans.

Lack of overview

New exploration initiatives are multiplying faster than ever before. Researchers working in the four corners of the world are discovering, describing, publishing, and collaborating—all the while generating a phenomenal quantity of data. Science is being written at a dizzying pace. How is it possible for the interested general public, or even the scientists themselves, to draw a clear and synthetic vision from this tsunami of information? An Internet search on certain deep-sea ecosystems, like the submarine canyons or ocean trenches, shows how little knowledge we have fished from the depths to date. Some very detailed but fragmentary data can be gleaned, though no amateur could make any sense

out of these, let alone integrate them into a coherent whole. Some habitats, like hydrothermal vents, have had the benefit of good media coverage ever since their discovery, and there are many accessible Web sites for these. But this is not the case for other biotopes found among the dark layers of the oceans: it was high time to ask researchers and scientists who have personally participated in the great conquest of the deep to assemble for us the pieces of this giant puzzle. Their texts enable us to appreciate the diversity of the deep-sea habitats, whether pelagic or benthic, and to comprehend that the immense expanse that occupies the majority of our planet can be punctuated by landscapes and ecosystems just as different from one another as the coral reefs are from the polar icecaps, or the Costa Rican rainforest from the Arizona desert.

Sooner destroyed than understood

The deep sea is far from being the inert void it was once considered. We now know, at this critical time when researchers are not the only ones with the means of accessing the great depths, that the abyss bears the wealth of a thousand resources: stocks of fish with fine and highly-prized flesh, deposits of ores, diamonds, and hydrocarbons, species with promise for medical or industrial research, and so forth.

With an exploding human population that exerts a disproportionate pressure on the natural resources at the surface, the deep sea represents huge global stakes, attracting the attention of various nations and great economic agents. The exploitation of deep-sea resources is no longer a theoretical possibility, but rather a tangible reality that has already caused irreversible damage. The most obvious victims of this are the cold water reefs extending between 200 and 2000 m depth, which shelter fish of a high commercial value. The massive decline of fish populations near the surface (certain species like tuna or swordfish have fallen by 90% in the last fifty years) have propelled industrial trawlers to scrape deeper and deeper into the substrate in order to find the catch that previously was readily available in shallow water. Deep-sea trawling is unanimously recognized as the most destructive of all fishing methods: the gigantic, weighted nets bulldoze the seafloor, leaving a wake of devastation in

exploitation is not likely to stop any time soon. The freedom granted by the United Nations for all to navigate, fly over, place cables in, fish from, and perform scientific research on the high seas results in irresponsible exploitation of fragile resources that are much slower to renew themselves than their surface counterparts. When the technology to exploit the depths was limited, this freedom posed no immediate danger. But today, it must be recognized for what it is: a legal loophole that allows, if not encourages, environmental destruction. A significant portion of our planet is being ravaged by a handful of humans, and this is occurring without the knowledge of the greater public. How is this possible?

To recognize, understand, and learn how to protect a new ecosystem requires much more time and attention than it does to exploit and destroy one.

"*Homo sapiens* is poised to become the greatest catastrophic agent since a giant asteroid collided with the Earth 65 million years ago, wiping out half the world's species in a geological instant." Dr. Richard Leakey

the underwater landscape. Obviously, fishermen do not destroy the reefs with malicious intent, but the result of this technique is that deepwater coral expanses, which are between 4000 and 10,000 years old (and not even properly indexed), have been disappearing faster than they can be studied and understood. Already half of the coral reefs off Norway have disappeared as a result of the deep trawling that began no more than twenty years ago. Clearly such practices are unsustainable, but because the high seas are the least protected zones of the Earth, the

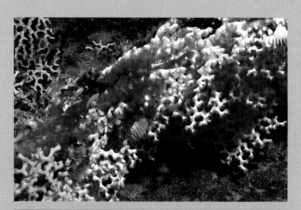

As early as 1934, deep-sea pioneers William Beebe [LEFT, TOP PHOTO] and Otis Barton [RIGHT] reached a depth of 900 m and sounded the ocean darkness on board a minuscule steel sphere called the bathysphere [TOP PHOTO], suspended from a hazardous cable. In 1960, the bathyscaphe *Trieste* [SECOND PHOTO]—a small pressure-resistant sphere topped by an enormous fuel tank—designed by the Swiss professor Auguste Piccard, made the deepest dive ever in history, at 10,916 m, in the Mariana Trench. The bathyscaphe was a sort of deep-sea elevator: it could go up and down, but could not move laterally. Piccard's engine nonetheless opened the way to free navigation in the deep and inspired the development of more sophisticated submersibles, like the American *Alvin* [OPPOSITE] or the French *Archimède*, which performed about eighty dives

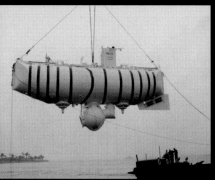

TOOLS FOR EXPLORATION

to over 9000 m depth. Thanks to its ability to reach depths of 6000 m, the French submersible *Nautile* [THIRD PHOTO] has access to 97% of the world's oceans.

Because it has a large Plexiglas sphere, which offers exceptional views and room to allow observers to sit comfortably, the *Johnson Sea Link* [FOURTH PHOTO] is an exception among submersibles. They are usually narrow, very cold units, with a limited power supply, and obviously without sanitary systems.

These many factors materially limit the duration of human immersion and lead researchers to prefer robots, which are maneuvered from support vessels. These ROVs (remotely operated vehicles) are able to do dives lasting several days because of the cable powering them from the surface. They are equipped with cameras that transmit footage in real time, as well as with mechanical arms that can gather samples from the depths. *Kaiko* [FIFTH PHOTO], belonging to the Japan Agency for Marine-Earth Science and Technology (JAMSTEC), was the world's deepest diving ROV. It was able to reach depths of over 10,000 m, but it was lost at sea in 2003.

The most recently developed of the exploration machines, the AUVs (autonomous underwater vehicles), like the one belonging to the Monterey Bay Aquarium Research Institute [BOTTOM PHOTO], are no longer constrained by cables and can freely crisscross the ocean bottoms. Though they do not yet have the extensive range of functionalities of ROVs, they are well adapted to certain disciplines like chemistry and physics, because they can be programmed to make various sorts of analyses of the water

OPPOSITE
Computer-generated image
Alvin, the American star of the submersibles, explores the gigantic chimneys of the Atlantic Ocean site known as "the Lost City." These enormous stalagmites are not colonized by any of the usual hydrothermal vent animals. At 60 m in height, they are the tallest hydrothermal edifices ever discovered. Geologist Jeff Karson noted that "if this vent

> "From the point of view of evolution, man is a success. But the most remarkable success of all this evolution, isn't it precisely the extraordinary cycle that leads him back to his origins, to the depths of the seas for which his blood is still nostalgic?" Jacques Piccard

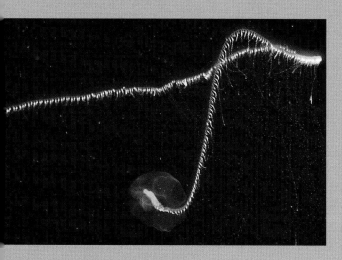

If tropical reefs were exploited as are those in the deep, the public outcry would have pressured governments to cease that sort of activity. In the case of the deep sea, only a few specialists are properly equipped to understand the full measure of the terrible process. Public opinion cannot possibly develop until people are enlightened about the exceptional natural heritage existing at the bottom of the seas. Our first duty is to *know* the world down below, and to be motivated by the wondrous biodiversity of our planet; not only the manifest beauty of the Blue Planet, illuminated by the sun for all to see, but also that of the hidden zone, the Black Planet, with its concealed beauty enshrouded beneath its tons of dark, impenetrable water.

SELF-MUTILATION

When one has the insane privilege of diving into the depths of the oceans on board a submersible to admire firsthand the strange and unsuspected creatures populating the liquid entrails of the Earth. it is impossible not to experience profound, primitive emotions that surprise the senses and stimulate the mind and touch

a fragile zone within, at once infantile and animal. Anyone who has had the chance to spend time in the nether realm of darkness has expressed, in one way or another, this shock that carries us back to our aquatic origins. One might mistrust the sincerity of these accounts, but once immersed several hundred meters below the surface, face to face with raw, untamed life, a truly primal emotion seizes hold of us. Yes, we are little pockets of moisture fighting against evaporation in a hostile, arid environment. Yes, the sea is our cradle, the pond that nourishes all life on Earth. A deep dive allows one to understand this on a level deeper than the intellectual. It's an experience that should be offered to every human being, a baptism as an adult that lets us renew our intimate connections with the chain of the living. To jeopardize its healthy existence is to risk losing contact with our identity and our future. It's deluding ourselves about our past and maiming ourselves for the future.

There is still time to pull back from the precipice of irreversible destruction, and protect humankind from a legacy of flagrant neglect and folly. •

LEFT
Praya dubia
Giant siphonophore
_SIZE up to 40 or 50 m long
DEPTH 700–1000 m

This siphonophore, a relative of the more familiar jellies, is the largest invertebrate on the planet, longer even than the blue whale, which is known for reaching lengths of 30 m. These fantastic animals explode into unrecognizable pieces when removed from their liquid element. *Praya dubia* has been known since the nineteenth century, but its incredible length was discovered only after the Monterey Bay Aquarium Research Institute (MBARI) undertook the systematic study of the water column in 1987.

OPPOSITE
Paraeuchaeta barbata
Predatory copepod
_SIZE 1.2 cm
DEPTH 200–1500 m

Copepods are small crustaceans that play a fundamental role in the oceans' food chain: they act as a link between phytoplankton, which they ingest in massive quantities, and other creatures, such as fish, jellies, and mammals. Copepods rarely represent less than 60% of the planktonic biomass, and in certain areas, their proportion even reaches 90%. They can easily be distinguished from shrimps because of their antennae, which form a right angle with their body, and by the fact that they hop about like fleas, in irregular jumps. This species of predatory copepod reproduces once per season and carries its eggs on its tail.

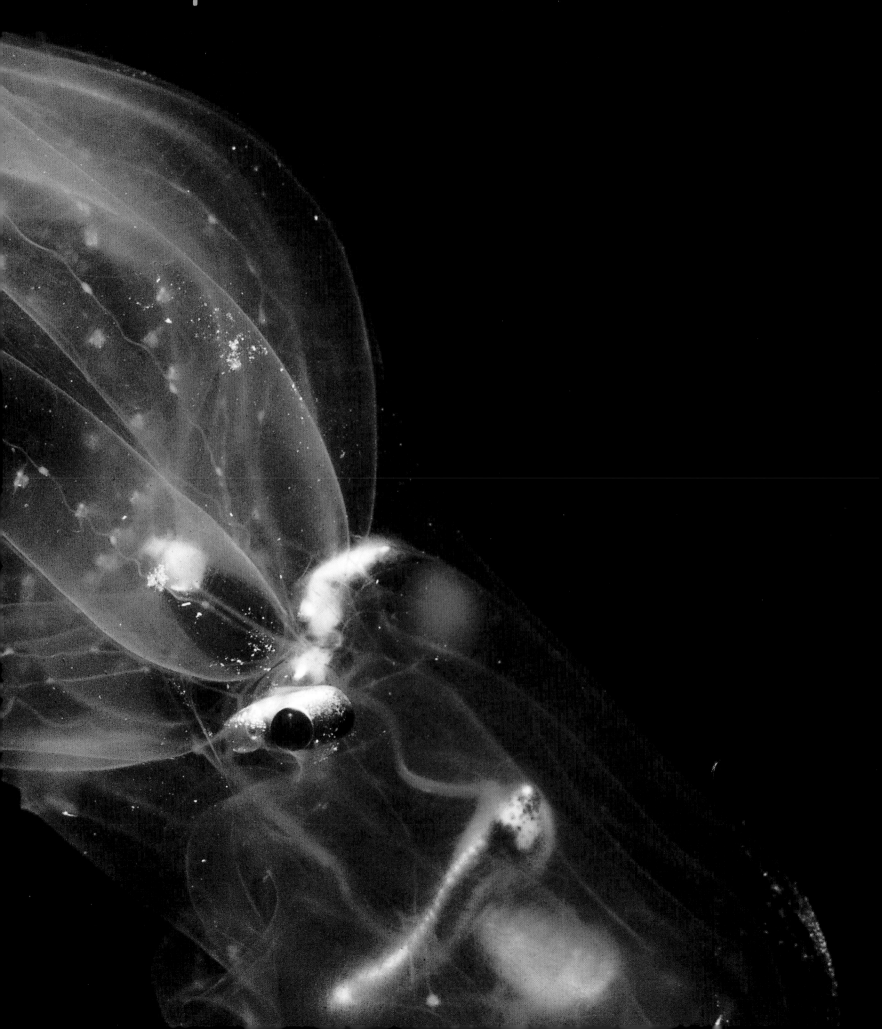

It is hard to imagine that the tremendous volume of water beneath the surface of the oceans can be "organized" in any manner. But the watery world follows rules that can be quite precise. These patterns have long remained mysterious, partly because the earliest submersibles crossed the water column without lights in order to save as much energy as possible while hovering over the bottom; the scientists on board took advantage of these long journeys through the darkness to doze. When midwater dives were finally performed during the 1980s, they found organisms that could represent as many as a million new

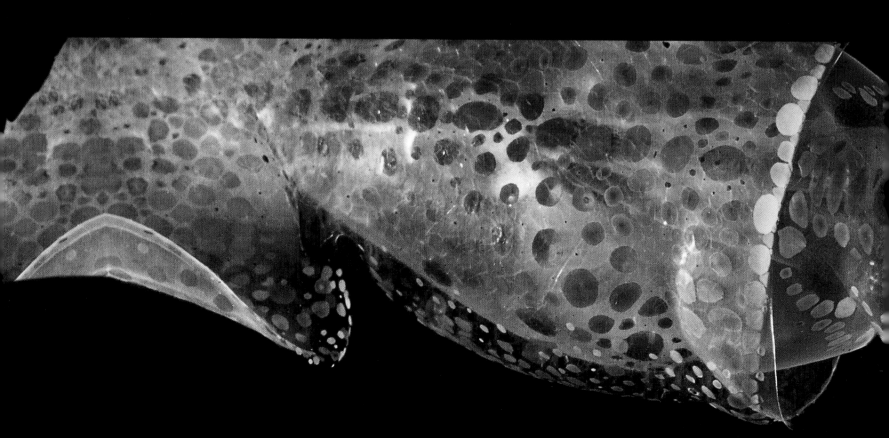

species. Each deep-sea dive still offers the possibility of

meeting some creature that no human has ever seen before.

The oceans have by no means revealed all their secrets,

even if researchers now know a great deal about the various

phenomena providing cadence to life in the heart of

this vast liquid space. Some of the ideas discussed here are

crucial to our new understandings about the largest realm

on the planet.

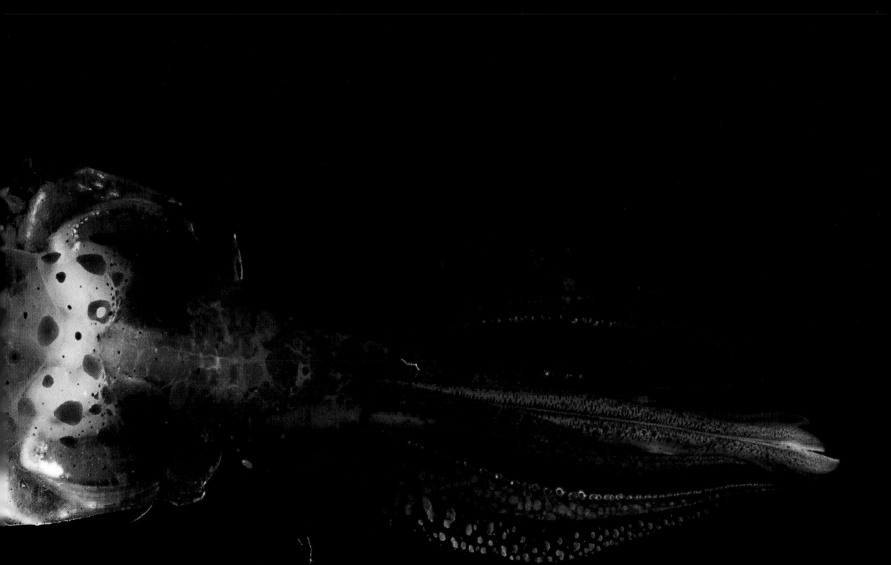

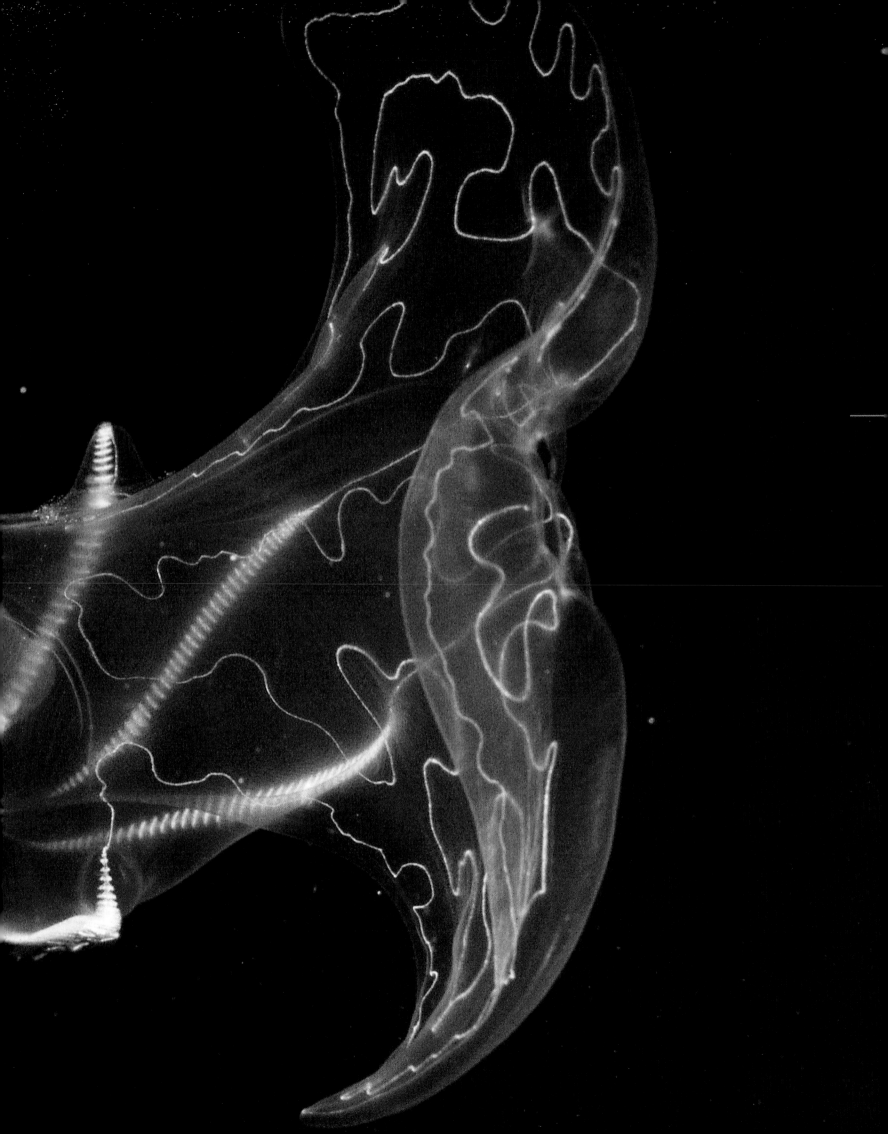

OPENING DOUBLE PAGE SPREAD

Vitreledonella richardi
Glass octopus

_SIZE 45 cm
DEPTH 200–2000 m

Most octopuses live on the seafloor, but some of them, like *Vitreledonella*, spend their entire life in the midwater. With no teeth, no poison, and no shell, they make a perfect prey for a number of the predators lurking in the water column—a good reason for them to try find a way to go unnoticed. The glass octopus, in fact, chooses the option of nearly perfect transparency, and only its opaque digestive gland can possibly betray its presence. Therefore, this cylindrical gland is always kept vertical, which minimizes its silhouette.

PAGES 30–31

Unidentified species
Planctoteuthis sp.

_SIZE about 20 cm
DEPTH about 1000–4000 m

There are five known species of this fragile and mysterious squid in the genus *Planctoteuthis*. Its body is covered with chromatophores that dilate and change color at the blink of an eye so that it always blends in with its environment.

PRECEDING DOUBLE PAGE SPREAD

Unidentified species
Llyria sp.

_SIZE 15 cm
DEPTH 500–3000 m

As is the case for many animals living at great depths, the biology, habitat, reproduction, and, even more so, lifespan of this creature remain largely enigmatic.

RIGHT

Helicocranchia pfefferi
Piglet squid

_SIZE up to 15 cm
DEPTH 400–1000 m

This small squid is easily distinguished by its characteristic muzzle—a siphon used for jet propulsion—which has the form of a pig snout. The young of *Helicocranchia* live close to the surface, descending further as they grow older in a process known as an ontogenetic migration.

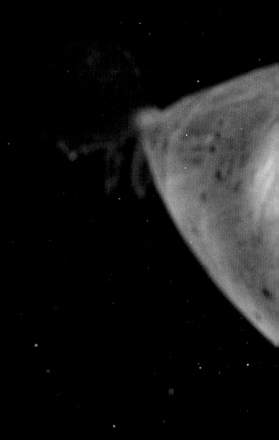

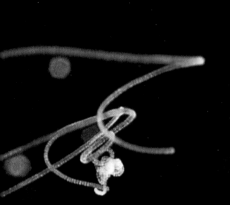

LEFT
Aglantha sp.
_SIZE 2 cm
DEPTH 320 m

The study of jellies and the other planktonic invertebrates is a very recent discipline, because these fragile animals disintegrate as soon as they come into contact with researchers' trawls. The great ocean depths shelter many more species still waiting to be discovered.

THE EXPLORATION
OF THE DEEP

Dr. Cindy Lee Van Dover
College of William & Mary, USA

Dr. Cindy Lee Van Dover
College of William & Mary, USA

OPPOSITE
Chauliodus macouni
Pacific viperfish
_SIZE 25 cm
DEPTH 250–4390 m

The meager food sources available at great depths have led species to extreme specialization, something that is not always advantageous. The viperfish has developed long and pointy fangs that leave little chance of escape for a prey. On the other hand, these teeth are so prominent that they do not even fit within the mouth of the viperfish, which has to put up with carrying its teeth around, outside its mouth, dangerously close to its eyes. If it miscalculates the size of its prey and impales an animal that is too big, it finds itself in the unpleasant situation of being unable to spit it out or to swallow it: condemned, therefore, to die along with its last meal.

In the first century A.D., Roman naturalist and historian Pliny the Elder believed that already the sea was understood, that the definitive list of marine fauna was complete—totaling 176 species!—and that, "by Hercules, in the ocean ...nothing exists which is unknown for us." Sailors of his time knew that our blue planet was covered with a skin of seawater over much of its surface, but they could not know the vastness of the volume of water beneath the surface, nor how many different creatures might live there. It wasn't until French mathematician Pierre Simon Laplace calculated the depth of the Atlantic Ocean in the late 1700s that we began to understand what the "deep" in "deep sea" means. At an average depth of 2.2 miles, the deep sea, the largest ecosystem on our planet, has been hidden from our view, inaccessible and unknown, for nearly as long as man has sailed upon it.

The deep sea was long perceived as a lifeless world. In 1858, British naturalist Edward Forbes wrote that life could not exist below 300 fathoms (1/3 mile). Forbes's "azoic theory" was soon thoroughly discredited by Sir Charles Wyville Thomson, who led the first oceanographic circumnavigation of the world: the Challenger Expedition of the 1870s. Over four years, Thomson and his colleagues scraped the seafloor with trawls and dredges at depths of up to nearly five miles and recovered more than 4000 new species of marine life. The dredged-up animals were often mangled almost beyond recognition, but they were nevertheless precious specimens that revealed hitherto untold tales about the rich diversity of deep-sea fauna. There were limits to what could be inferred from these samples; they often provided little insight into the way life on the seafloor looked, or into how the animals might interact with one another. To paraphrase explorer and humanist Théodore Monod, attempting to understand life in the deep sea using dredges is like aliens trying to understand life on Earth by blindly dangling a hook from space and retrieving a cockroach, a t-shirt, and an iPod. Trawls and dredges allow us to measure the biological diversity found in the deep sea—they are still used today for species counts and other statistics—but they are almost useless for understanding animal behavior in natural settings. To achieve this goal, one needs to observe organisms in their environment.

In the late nineteenth century, an underwater voyage was the dream of many adventurers inspired by Jules Verne's *20,000 Leagues under the Sea*, but it was not until the 1930s that the first explorers descended beyond where light penetrates, into the relentless dark, the veritable deep sea. William Beebe—lanky, literary, lyrical naturalist of the Bronx Zoo—was the leader of these first deep dives, ultimately making a round trip half a mile down. Otis Barton, a young man of large fortune, designed and built the bathysphere, a tethered metal sphere with an inside diameter of less than three feet in which the deep-sea pioneers cramped themselves for several hours during each immersion. In accounts of his dives, Beebe gives attention to the pale green "dancing" lights—the bioluminescent lanterns of creatures unnamed and never seen before by any man—which came into focus before his astonished eyes.

As Beebe explored beneath the surface of the sea in his bathysphere, Swiss scientist Auguste Piccard was making the first flights into the stratosphere, "out far beyond the atmosphere," nearly ten miles above the ground. To accomplish this feat, Piccard designed a pressurized, spherical gondola suspended beneath a hydrogen-filled balloon. Using the principles of design he learned from its construction, Piccard worked to fulfill his own dream of descending, untethered, into the depths of the sea. He built a small metal sphere that could withstand pressure and coupled it to a buoyant "balloon" filled with gasoline. Dream became reality when, in 1954, Piccard descended to a depth of 4000 m in his bathyscaphe, the first untethered class of vehicle to take people into the deep sea. In 1960 the *Trieste*, a second-generation bathyscaphe operated by Piccard's son Jacques and U.S. Navy lieutenant Don Walsh, descended seven miles to the deepest part of the ocean, the Mariana Trench. The Walsh and Piccard dive was more a record breaker than a dive of exploration, but it is a record that remains unmatched today: more men have walked on the moon than have dived to the deepest part of our oceans.

The technological successes of the bathyscaphes inspired a team of U.S. oceanographers led by geologist Al Vine to call for a smaller, more maneuverable submersible that could be used to explore the deep sea. With its gasoline balloon, the *Trieste* was inherently buoyant; she sank only when loaded with expendable weights. Thus she could descend and ascend, but she could not adjust her depth once her weights were dropped, nor could she move laterally. *Alvin*, the three-person submersible named for Al Vine, was the first deep-diving submersible to require a pilot who could drive over the seafloor by controlling the angle and speed of a large aft propeller. *Alvin* made its first dives in 1964, marking the commencement of the true age of ocean exploration.

Alvin, together with the new French submersible *Cyana*, demonstrated the merit of submersibles as scientific workhorses during an unprecedented mission of exploration: Project FAMOUS (French-American Mid-Ocean Undersea Study; 1972). Geologists were able to dive up to 2.5 miles below the surface and observe for the first time the Mid-Atlantic Ridge, the long volcanic mountain chain that bisects the Atlantic

Ocean. In the mid 1970s, geologists shifted their focus from the Atlantic to the Pacific, diving to 1.5 miles on the Galápagos Rift, where they encountered warm water (20°C or more) flowing out of cracks in the rocky seafloor. Soon after, they discovered spectacular hot springs (350°C) spewing from tall mineral chimneys on the East Pacific Rise, the mountain range that begins in the Gulf of California and extends southward off the coasts of Central and South America.

Geologists had predicted that hot springs, "hydrothermal vents," would exist on the seafloor, but no one anticipated the extraordinary communities of strange animals bathed in the flow of warm water. Reports of six-foot-long red-plumed worms living on chemicals in the water hastened return trips to the dive sites by biologists. The seafloor observations of the late 1970s motivated the development of deep-submergence assets by other nations. *Alvin* and *Cyana* were joined by other deep-diving research submersibles operated by French, Canadian, Russian, and Japanese teams.

Since the discovery of hydrothermal vents in 1977, the pace of exploration in the deep sea has steadily increased, fueled by the finding of novel adaptations to extreme environments and by the gain of fundamental insights into how our planet works. Our increasing ability to access the seafloor with new tools and sensors promotes and enhances exploratory activities. Tethered and untethered robots are now the tools of choice for many of the challenges faced by deep-sea explorers. Nevertheless, the construction of two new human-occupied submersibles, one Chinese and the other American, underscores the anticipated need for a human presence on the seafloor for the next half century.

Man has observed less than 1% of the seafloor; the challenge lies before us. During the twentieth century, the deep sea became accessible. In this twenty-first century, the deep sea will become known. •

OPPOSITE
Ridgeia piscesae
_SIZE almost 2 m
DEPTH 1800–2800 m

When hydrothermal vents were discovered, researchers found colonies of animals such as the red-plumed tubeworms that were surprisingly dense compared to those in the abyssal plains. The *Ridgeia piscesae* shown here are mixed with another, smaller species, the palm worms (*Paralvinella palmiformis*). Sometimes they cover great expanses, and this has inspired scientists to name the particularly extensive colonies "the Strawberry Field" and "the Rose Garden."

NEXT DOUBLE PAGE SPREAD
Nausithoe rubra
_SIZE 10 cm
DEPTH 600–2500 m

This jelly suspended in space is not found above 600 or 700 m of depth. Its velvety texture and very dark color provide it with a perfect camouflage that absorbs the bioluminescent glimmerings of the abyssal fauna.

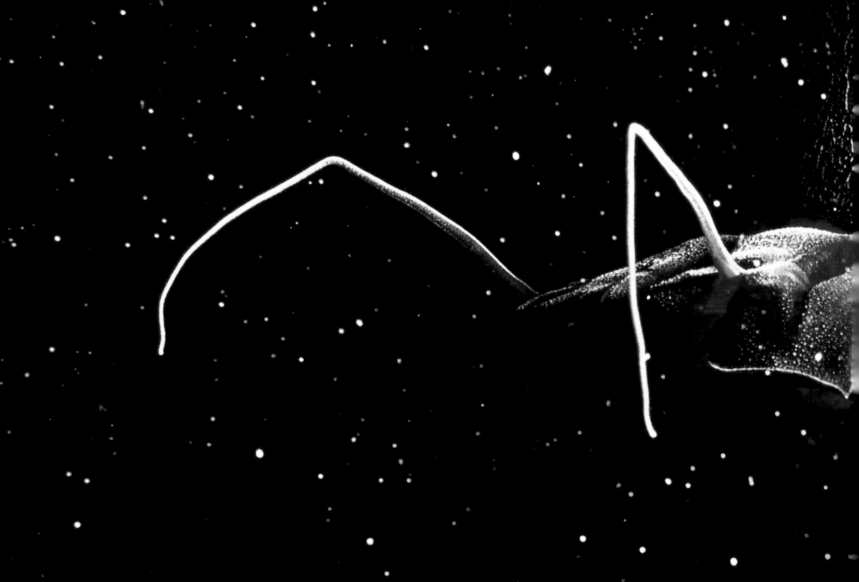

OPPOSITE
Histioteuthis heteropsis
Jewel squid
_SIZE 20 cm
DEPTH 400–1200 m during the day,
0–400 m at night

The jewel squid's body is completely covered with photophores, which make it look somewhat like a giant strawberry, as biologist James Hunt has noted. Depending on the light level of its immediate environment, it can turn its lights on or off to hide from predators as well as from prey. Despite the sophistication of its counterillumination routine, the jewel squid appears regularly on the sperm whale's menu: over two thousand *Histioteuthis* beaks were found in the stomach of just one of these cetaceans.

NEXT DOUBLE PAGE SPREAD
Paraliparis copei copei
Blacksnout seasnail
_SIZE 17 cm
DEPTH 200–1692 m

With its puppy-like head and its tadpole shape, this fish owes its name to the gelatinous substance that covers its scaleless body, though it is much more captivating than the animal to which its name alludes. When surprised or threatened, the blacksnout seasnail has been observed rolling itself into a loop. Scientists believe this may be a defensive posture through which it takes on the appearance of a jelly. In the darkness, an elongated form will bring to hunters' minds the edible shape of an eel, while a round ball will recall a stinging jelly that should be avoided.

"For a primate, the midwater is really a bit unsettling:

so much water, so much darkness, and in every direction...

The bottom is reassuring, even if it lies 4000 m

below the surface."

Théodore Monod, 1954

AT 150 M DEPTH, 99% OF SUNLIGHT HAS BEEN ABSORBED

BY WATER. BELOW 1000 m, IT'S TOTAL, INKY BLACKNESS FOR ALL.

MIDWATER LIFE: SURVIVAL IN A HARSH ENVIRONMENT

Dr. George I. Matsumoto
Monterey Bay Aquarium Research
Institute (MBARI), USA

OPPOSITE

Stigmatoteuthis arcturi
Arcturus jewel squid
_SIZE 30 cm
DEPTH 400–1200 m during the day,
0–400 m at night

If there is any creature in the depths
of the globe we can refer to as
eminently bizarre, it is without
doubt this jewel squid. One of its
two eyes is small and embedded
within its body, while the other
is extremely large and literally
protrudes out of its socket.
This animal is highly adapted to
living in the oceans' twilight zones;
its first eye is always directed
downward toward the darker layers,
while the second is oriented toward
the sunlit surface. In order to put
its specialized features to greatest
advantage, it is believed that
the squid maintains a 45° angle
as it swims.

The midwater of the ocean is the largest habitat on Earth. There is only one ocean, as all the large bodies of water are connected, and the midwater realm within this ocean stretches from the surface to the bottom—a vast, three-dimensional, fluid environment that presents some extreme challenges for all inhabitants. Despite the sunlight that disappears quickly as you descend, the bone-chilling temperatures of the depths (4–6°C), the low oxygen levels, and the ever-increasing pressure, there are untold billions of organisms inhabiting this immense water column. We find representatives from almost every animal phylum living in the midwater, and all of them face the same four basic challenges: staying in the midwater, finding food, avoiding predators, and finding mates.

The first challenge is to stay in the midwater. Floating up to the surface or sinking to the bottom often means death for midwater organisms that are used to a life without walls—and the ocean surface and bottom are dangerous walls. Even coming close to shore creates problems for many of the gelatinous animals, as the corals, rocks, wave action, and surge can tear these delicate organisms apart.

Some organisms actively swim to maintain their position or to move around in the water column. Other larger animals like fishes and some gelatinous zooplankton use air bladders to move up and down in the water column. Certain siphonophores use carbon monoxide in their air bladders (called pneumatophores) to regulate their movement in the water—adding more gas to move higher and removing gas to sink.

Sharks rely on a liver that contains oils, which provide lift by making the shark slightly less dense than seawater. In certain species, the oil in the liver is so efficient that it can reduce the density of the shark by 99%, greatly increasing its buoyancy.

Like all other animals, midwater organisms must find food in order to survive. As you go deeper into the ocean, food becomes scarcer; so the first challenge is to locate the food, the second is to capture what food is available. There are many methods of finding food—there are those active predators that roam the ocean looking for slower moving prey, ambush predators that set traps or agitate a bioluminescent fishing lure, and filter feeders that strain the water for food.

In the deep dark waters where most of the food available is what drops from above, animals often have small to medium-sized flabby bodies that require low levels of energy and allow them to fast for long periods of time. The gulper eel (*Saccopharynx*) and the swallower (*Chiasmodon*) can go weeks without a meal but have expandable stomachs that enable them to swallow whatever prey they encounter—even if it's as large as themselves. Fishes like the fangtooth (*Anoplogaster*) and the viperfish (*Chauliodus*) have a mouthful of sharp teeth that make sure they won't miss the catch.

Feeding, although an essential behavior, is not a well-understood phenomenon in the midwater. Much of our knowledge comes from gut content analysis, but the increasing use of ROVs (remotely operated vehicles) in recent years has permitted scientists to explain some perplexing questions that originated from the very first dives of William Beebe in the 1930s; they have also generated a whole new suite of questions that we are still trying to answer. The barreleye spookfish (*Macropinna*), with upward-looking tubular eyes, a mouth in front, and a transparent head, is an excellent example. Direct in situ observations have now revealed that this marvelous fish locates food using the dim downwelling light from the surface and then rotates its eyes from upward-looking to forward-looking in order to keep its eyes on the prey before opening its mouth.

One of the new, not yet answered questions focuses on what these animals do in terms of sleep—Do they? Do they ever rest?

And as if maintaining position and finding food weren't difficult enough, organisms also need to reproduce. In this enormous three-dimensional habitat, it seems impossible that animals can find a mate, and yet they do. There are hermaphroditic animals like the comb jellies that are able to reproduce on their own, and some squid species that mate when they encounter each other and then store the sperm until some later date when the eggs are ready to be fertilized. The difficulty with locating a mate is that often the devices used for interspecies communication can also be used by predators. Imagine being a lanternfish (Myctophid) trying to signal to another lanternfish that you are available and looking for a mate—but a viperfish lurking in the darkness uses the light that you are making to target you. Sometimes those cunning predators can even mimic their prey's light signals to lure unsuspecting victims who think they have found a mate!

All of these reproduction strategies are useless if you end up being somebody else's meal, so another priority is to avoid predators. The midwater offers very little in the way of places to hide: unlike land, there are no rocks or trees to hide behind. Animals can try to avoid

danger by various means: they can either be very small and difficult to find or large enough to deter predators; that is the option chosen by the longest animal in the world, the siphonophore *Praya*, which can reach up to 40 m in length with over 300 stomachs! Animals can also be transparent or incorporate some type of camouflage so that predators can't see them. The transparent octopus *Vitreledonella* stalks the water looking for prey while avoiding predators by being almost completely transparent. The squid *Galiteuthis* uses light organs to mask the eyes so that as the squid turns, the light organs rotate to always provide masking from predators below.

Another good survival strategy is to live in an environment that is unsuitable to most predators, as does the much maligned vampire squid (*Vampyroteuthis infernalis*). This creature is not a blood-sucking parasite, but a beautiful deep red cephalopod that lives in the oxygen minimum layer. This layer may be as thick as 300 m in some regions of the ocean and offers protection from predators that require higher levels of oxygen to survive.

The responses of animals to the harsh three-dimensional habitat called the midwater are extremely diverse and certainly still poorly documented. What has been made extremely clear over the past few decades is that as much as we are learning about the midwater realm, there is so much more to find out. New species are constantly being described, and we are finding that we really cannot hope to understand the ecosystem until we know who the "players" are and what role each species might play. •

Melanocetus johnsoni
Black-devil anglerfish
_SIZE 20 cm
DEPTH 1000–4000 m

This female with her flabby and extendible skin has all the characteristics of a ferocious sea monster, but she is no larger than a grapefruit! The round bodies of anglerfish are adapted to their stationary lifestyle. Besides saving energy, this absence of movement is also an effective way of avoiding detection by predators or by prey who are sensitive to even the slightest vibrations of the water. The males are dwarves that remain attached to a single female for life and gradually dissolve their tissues into hers, eventually disappearing completely.

NEXT DOUBLE PAGE SPREAD

Winteria telescopa
Spookfish
_SIZE 20 cm
DEPTH 400–2500 m

The spookfish's two enormous telescopic eyes are perfectly adapted for collecting the maximum available light. They provide the fish with an uncontested visual advantage for hunting in the dark. The large number of rod cells in its retina allows it to distinguish bioluminescent glows from ambient light and thus to avoid a significant number of the traps that creatures of the deep lay for one another.

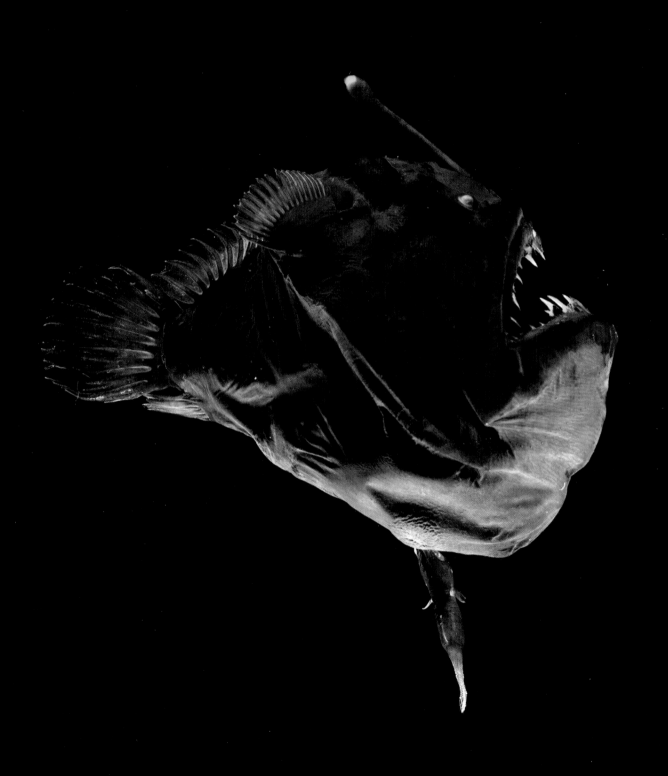

the midwater is a vast, open jungle affording no possibilit
from any direction: the sides

OF SECLUSION. DANGER IS CONSTANT AND CAN COME
THE FRONT, OR THE BACK, BUT ALSO FROM ABOVE OR BELOW.

Teuthowenia pellucida
Googly-eyed glass squid

_SIZE 20 cm
DEPTH larvae and juveniles 0–900 m,
adults 1600–2500 m

When alarmed by a passing predator,
a googly-eyed glass squid undergoes
an astonishing transformation: first,
this normally slender squid inflates
its body with water, swelling into a
transparent sphere. If this does not
discourage the threat, the squid
draws its head, arms, and tentacles
into its cavity. As a last resort,
the animal will fill its cavity with ink,
disappearing into the darkness.

RIGHT
Anoplogaster cornuta
Fangtooth

_SIZE 15 cm
DEPTH 600–5000 m

The fangtooth's menacing appearance
is accentuated by the skeleton-like
appearance of its strange protruding
bones. The compartments showing
through beneath its scanty flesh are
actually extremely sensitive sensory
canals that allow the animal to
detect even the slightest motion
in the water. In the depths where
the fangtooth lives, the blackness is
absolute, the cold and pressure are
extreme, and food sources are
scarce. With its disproportionately
large jaw and its razor-sharp fangs,
the fish is outfitted to confront the
severity of its environment.

NEXT DOUBLE PAGE SPREAD, LEFT
Stauroteuthis syrtensis
Glowing sucker octopus

_SIZE up to 50 cm
DEPTH 700–2500 m

The elasticity of this octopus's body
provides it with uncommon
capabilities for metamorphosis:
in the space of just a few seconds,
it can inflate, contract, elongate,
or twist itself.

NEXT DOUBLE PAGE SPREAD, RIGHT
Chaetopterus sp. nov.
Pigbutt worm

_SIZE 2 cm
DEPTH 965–1300 m

This pair of flying buttocks is a new
species of worm recently discovered
by MBARI scientists in California.
It is closely related to polychaete
worm species, which live in tubes
fixed to the ground and whose larvae
spend several months in the water
column. In adapting itself to a pelagic
lifestyle, this animal has probably
conserved some of the morphological
features characteristic of the larvae,
which would explain its very unusual
shape for a worm. The pigbutt
worm feeds by inflating a balloon
of mucus, which collects organic
particles on its surface. Only ten
specimens have been observed

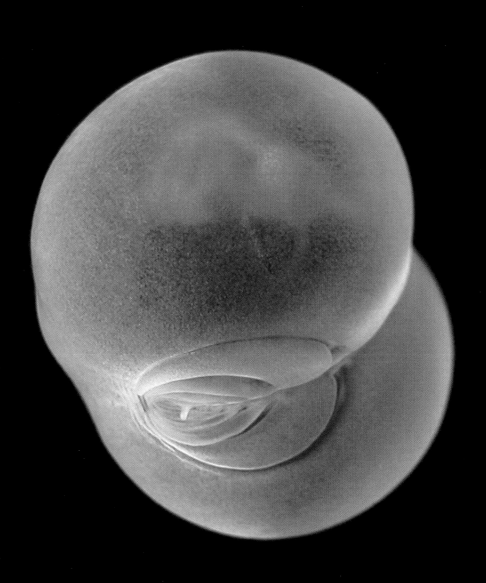

A CHINESE SHADOW THEATER

In order to get a sense of the life of the creatures in the first hundreds of meters in the oceans, where some light still penetrates, one needs to start by understanding what their environment is about. Since the surface above them is bright and the water underneath is dark, to any creature situated below them,

their presence stands out as clearly as a silhouette in a Chinese shadow theater. The animals that bear appropriate defensive weapons—large size, teeth, or poison—can afford the luxury of being noticed, but those that do not had best try to blend in with the background. Various options are available, among which transparency figures foremost; it is widely used among marine organisms. For the human eye, this is a perfect camouflage; even at just a few centimeters' distance, it is impossible for us to see many of the animals composed of translucent gelatinous matter, even if they span several meters. This "glass menagerie," as dubbed by biologist Sönke Johnsen, is composed of jellies, salps, comb

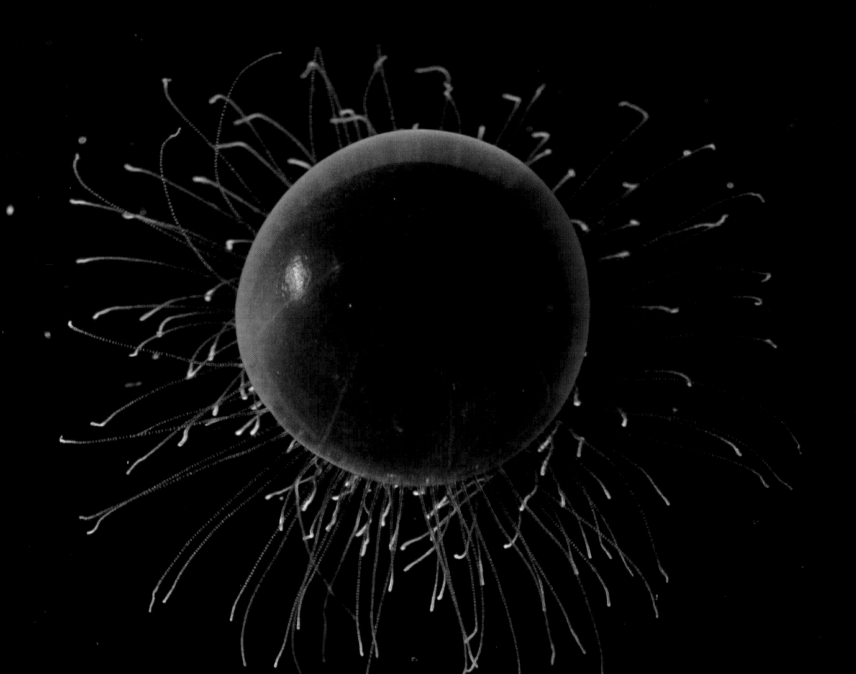

jellies, marine snails, and the like. If an animal cannot achieve transparency, it can resort to counterillumination through a battery of ventral photophores. These allow it to mask the opacity of its silhouette by adjusting the intensity of its light emission to match the level of ambient light. This is the tactic that was used by airplanes during World War II, but well before that, it had been adopted by various deep-sea fishes; the best known example is the silver hatchetfish (*Argyropelecus* sp.).

If several hundreds or thousands of creatures have independently developed sophisticated camouflaging techniques,

then doubtless the reason is that they work. But in the mad race for the most creative invention, the predators are not to be outdone, and some of them have already learned how to thwart the tricks of their prey. Certain squids, for example, are able to detect polarized light (as though they wore the equivalent of our polarized sunglasses) and thus of distinguishing creatures whose bodies allow 90% of available light to pass through. Others have their eyes covered with a yellowish filter, which allows them to make out, in greenish highlighting effects, the photophores of fish deploying the most marvelous effects of counterillumination. These innovative responses to the adaptations of prey remind us that evolution remains an ongoing process. •

OPPOSITE BOTTOM AND BELOW
Vampyrocrossota childressi
Black medusa
_SIZE 1.5 cm
DEPTH 600–1500 m

61

LIFE IN THE WATER COLUMN

Like a black hole in space, this jelly absorbs all the light that strikes it. One needs to carefully watch the black medusa in order to discern that it is not simply composed of translucent gelatin, but that it also possesses a velvety, dark umbrella. This creature has achieved absolute perfection in the art of camouflage: it is completely invisible in its dark environment, which could explain why it was discovered only in 1992 in the Monterey Canyon.

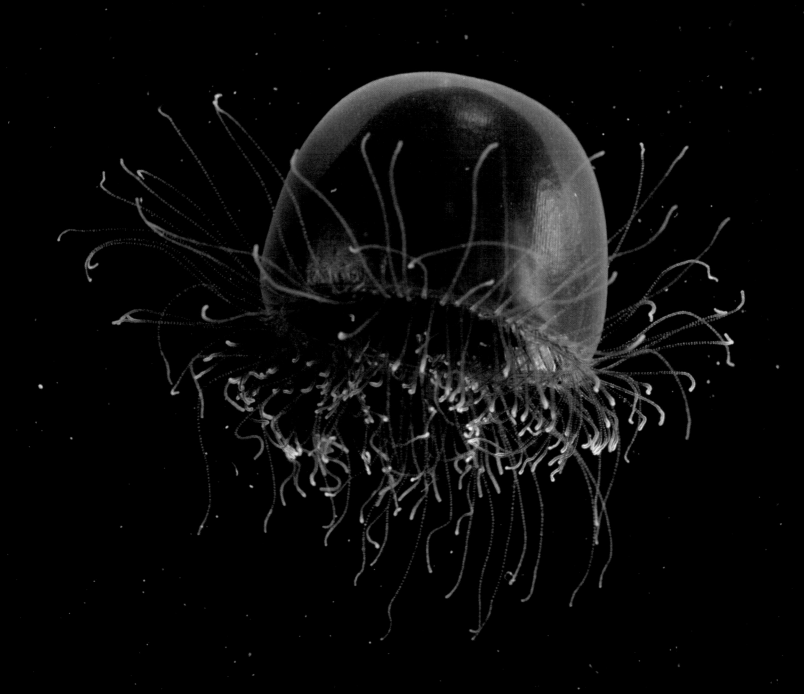

Leachia sp.

Bathyscaphoid squid

_SIZE 15 cm
DEPTH 730 m

Cranchiid squids are inoffensive soft-bodied creatures with no stings and no defensive armoring devices or spikes, and that, based on the testimony of their numerous predators, must be exquisitely delicious. These vulnerable animals are thus faced with no other choice than to try to pass unnoticed, and accordingly they have opted for an almost complete transparency. The bathyscaphoid squid was named in honor of the submersible invented by Auguste Piccard, because the creature, like the vehicle, reserves an enormous space—the two transparent internal pockets—for its buoyancy chamber. The sacs are filled with ammonium ions, which are lighter than water, allowing the squid to achieve neutral buoyancy.

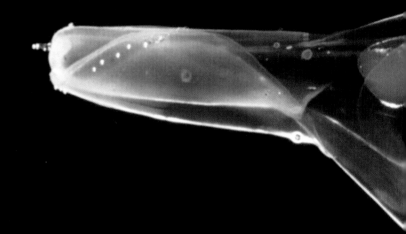

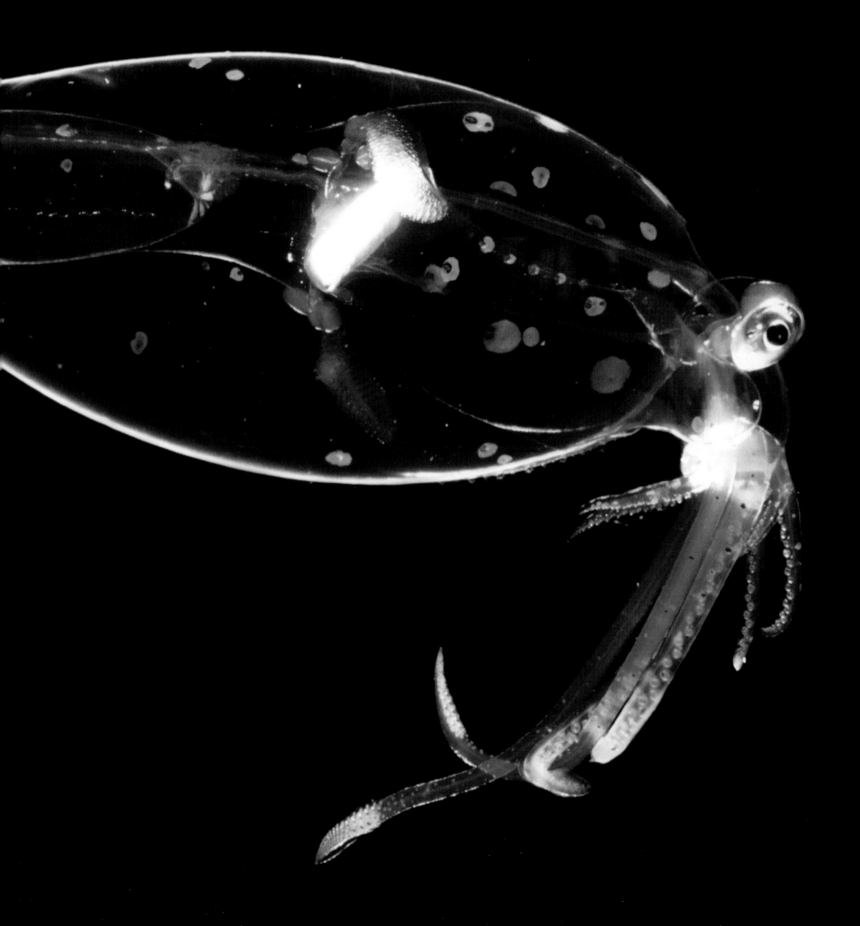

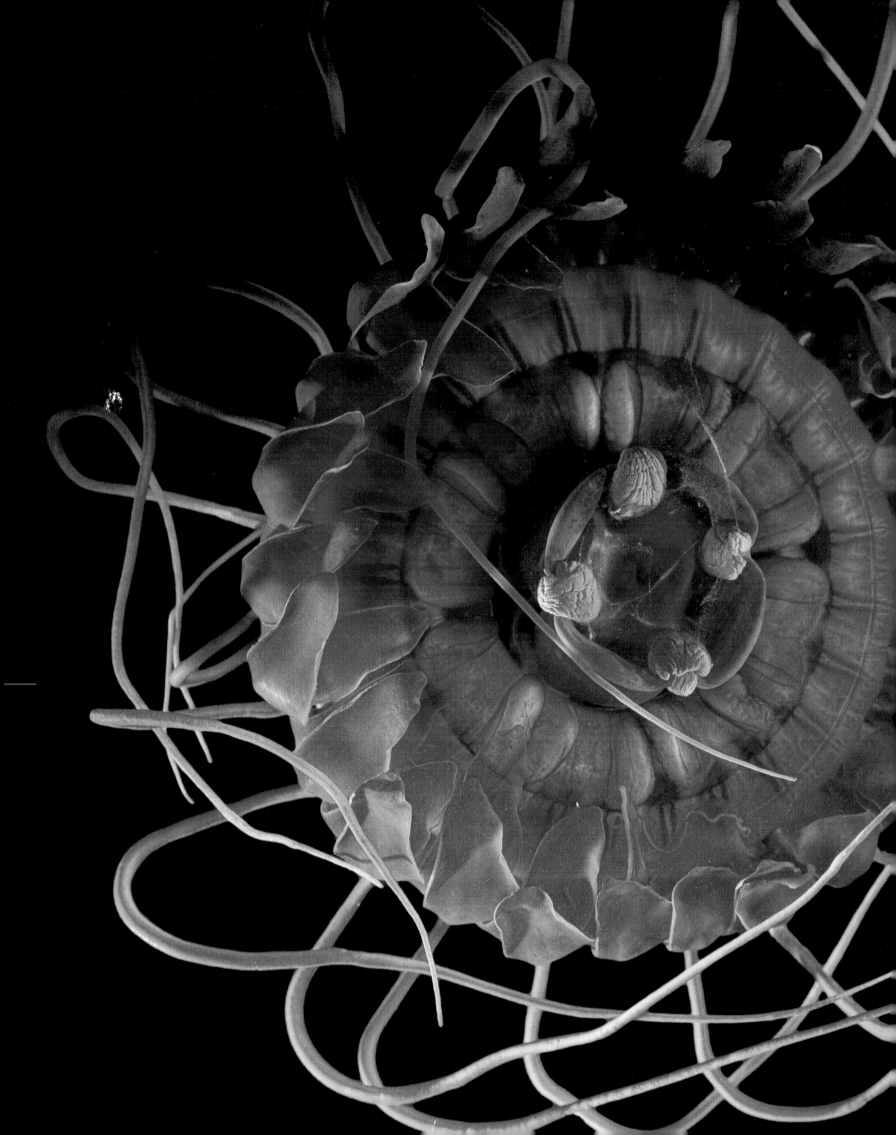

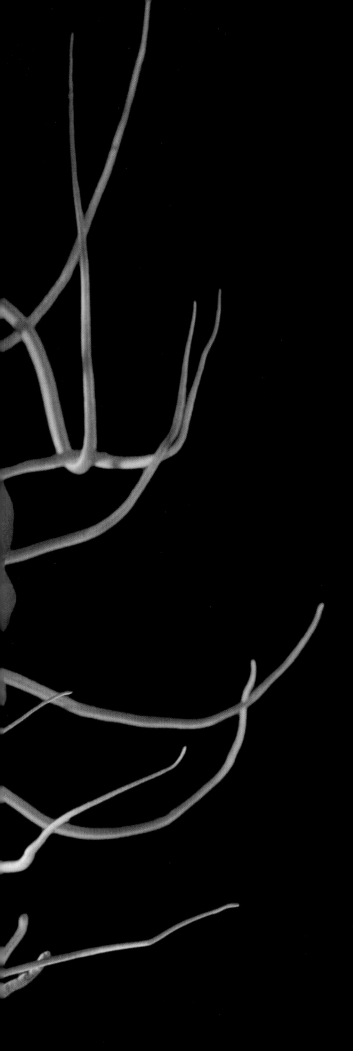

LEFT
Atolla wyvillei
_SIZE 15 cm
DEPTH 600–5000 m

The *Atolla* jelly, which is widespread at great depths, is known among explorers for its remarkable display of bioluminescence: a sort of rotary disk that suddenly lights up with thousands of blue flashes. *Atolla wyvillei* puts on this show when it is threatened and it wants to attract the attention of another predator to its attacker. This strategy corresponds basically to our own deafening burglar alarms, except that in the deep-sea realm, it is not sound, but light, that is most effective.

NEXT DOUBLE PAGE SPREAD, LEFT
Munnopsis sp.
_SIZE body 1–2 cm; legs 15 cm
DEPTH 900–3000 m

When seen in motion, one would think it was a spider playing an invisible harp... Strange and graceful, this animal is unique among the inhabitants of the water column; it has puzzled all the scientists who crossed its path so far. Only biologist Karen Osborn has studied these creatures alive, and what she's discovered about them makes them even more fascinating. These cousins to the woodlouse can swim in the water column, to more than 3000 m above the bottom, just as well as they traverse the substratum with their long legs. The incredible distances that these tiny creatures manage to cover speak volumes about the efficiency of their swim stroke. It's no surprise to learn that, with such long legs, the isopods have no other choice than to swim backward, dragging the elongated walking legs behind them. They paddle using their swimming legs as oars, which have bristles that increase their surface area on the downstroke. The isopods raise their young in a ventral pouch, as can be seen in the photo, which can hold hundreds of larvae. In all likelihood, the mother or pouchmates devour some of the young before they attain independence.

NEXT DOUBLE PAGE SPREAD, RIGHT
Larvae of Spantagoid heart urchins
_SIZE 2 cm
DEPTH unknown

With their right angles that are seldom found in nature, these giant larvae look like two spacecrafts. They may be the larvae of deep-sea urchins.

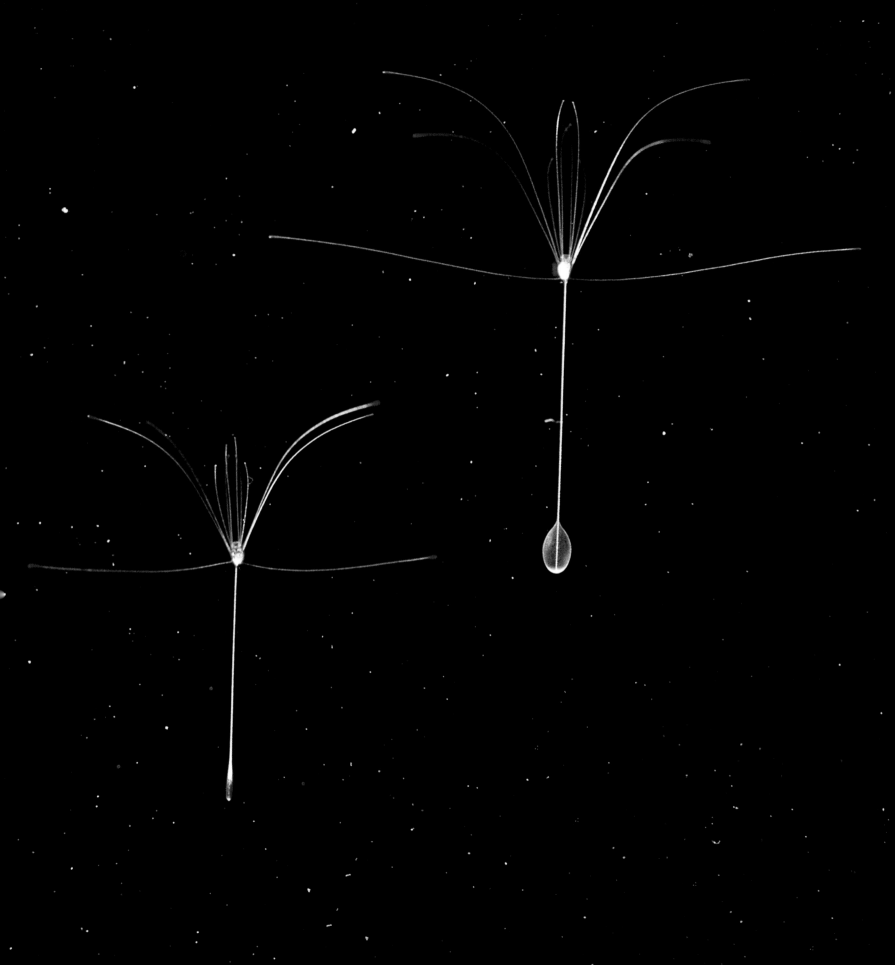

OPPOSITE
Crossota millsae

_SIZE about 3 cm
DEPTH 1000–3800 m

Although this graceful little jelly
is abundant in the North Pacific,
it was discovered only recently
because of the extreme depths at
which it lives. The creatures are not
found at all above 1000 m, and
the greatest densities occur around
2000 m. Its shimmering colors
and fascinating beauty inspired
the scientists who discovered it
to honor their colleague, biologist
Claudia Mills, by naming the jelly
after her.

"Give me a map, give me a ship, and I can take you

to where we are going to discover entirely new things."

Cindy Lee Van Dover, 2005

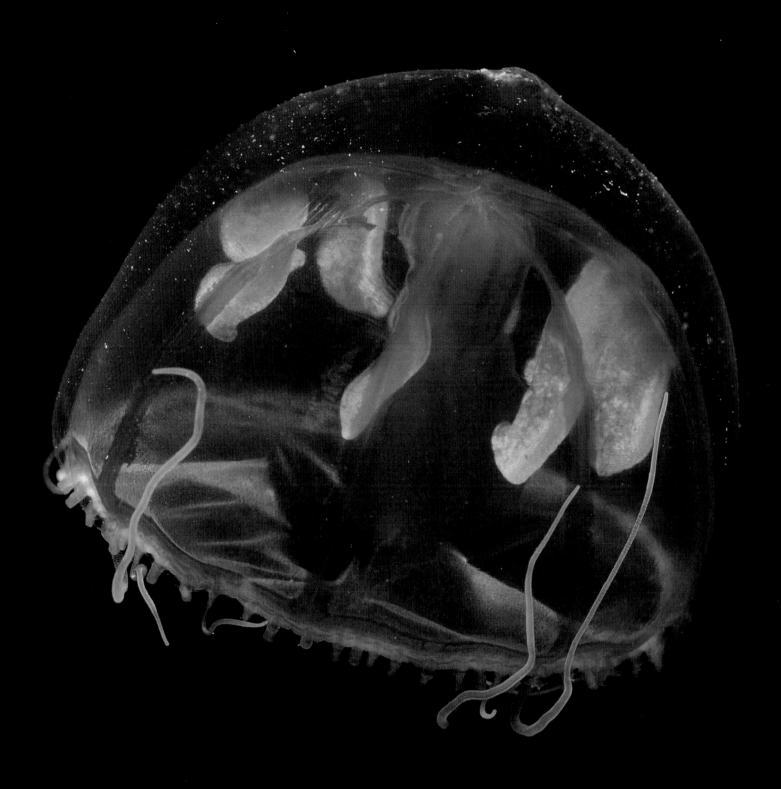

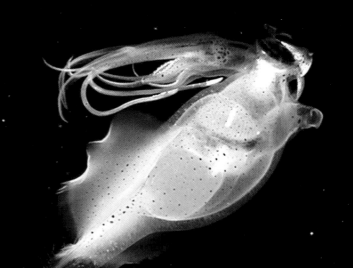

THE NOCTURNAL BALLET OF DEEP-SEA CREATURES

Dr. Marsh Youngbluth
Harbor Branch Oceanographic
Institution, USA

OPPOSITE
Planctoteuthis oligobessa
_SIZE 20 cm, including tail
DEPTH 1000–4000 m

This small squid is as rare as it is fragile. Only about a dozen of these creatures have been described because of the great difficulty in collecting them intact. The first in situ photos, showing their strange corkscrew-like tail, which breaks off when they are captured, surprised even the scientists. Their tails recall the screws of early steamships and suggest some locomotive purpose.

At dusk and at dawn on calm, clear days, the surface of the sea appears to be a huge stretch of impenetrable water. Nothing "outside" provides a hint about the frantic activity occurring "inside" the sea. Nevertheless, each evening and morning all the oceans and even the lakes of the world are a theater to mass movements involving billions and billions of creatures swimming from deep waters to the surface and then back again to a colder, darker world. When these animals migrate they often aggregate and form dense layers. Sixty years ago, when active sonar was first available, captains of fishing vessels thought that the bottom was rising under their boat. This phenomenon, called the "vertical migration," is the largest synchronized animal movement on Earth.

Marine biologists have known since the nineteenth century that many kinds of animals move from one depth to another each day. The most obvious demonstration of vertical migration is found by trawling nets at the sea surface. The catch of organisms is often much larger at night than during the day.

All kinds of creatures, large and small, migrate vertically in every ocean. They perform this ritual 365 days of the year. Like nomads in deserts, they move to and from the surface waters in search of an oasis: the first 100 m of the sea, where the sunlight still penetrates, the "photic" zone, abounding with food. Here, microscopic plants transform inorganic material such as carbon dioxide into sugars thanks to the sun's energy. Their performance serves as the basis of the whole oceanic food chain. Further

down in the dark deep sea, there is no more plant life to fuel this photosynthetic production, so virtually all life in the deep ocean depends on what is produced in the sunlit surface. Like wild animals everywhere, deepwater fauna migrate only when they must and only when it's reasonably safe, under the cover of twilight or darkness. The signal that triggers and resets this ritual is the daily waning and waxing of sunlight, so travel frequently starts at dusk and ends at dawn. The time spent in shallow waters is usually brief, lasting only a few hours.

Who are the central characters in this recurrent play about survival? Small crustaceans— 1–3 mm long—called copepods are the most ubiquitous and abundant migrants. Larger animals, mostly 10–30 cm long, such as jellyfishes, krill, squids, and fishes, are also common travelers. The smaller animals swim at a tortoise-like pace of just a meter or so per minute. At this rate several hours are needed to navigate to and from shallower water. The vertical speeds of larger fauna are much faster, ranging from 100 to 200 m per hour. Regardless of size, these vagabonds traverse daily distances of tens to hundreds of meters—perhaps as much as 1000 m for some species. Animals that live below 1000 m rarely undertake these extensive vertical migrations; the surface being so far, the time spent traveling is not worth the effort.

How do deep-sea animals perform this extraordinary effort? Copepods, often called the insects of the sea, have developed plumose and paddle-shaped appendages that enhance buoyancy and facilitate navigation in a watery environment that is eight hundred times as dense and fifty times as viscous as air. Jellyfishes with unfamiliar names, like medusa and salp, are surprisingly strong swimmers. Instead of pushing their way through the water, these animals move by jet propulsion, sucking in water and squirting it out over and over again to generate thrust. The most fragile gelatinous species, such as ctenophores and siphonophores, can alter the gases and chemicals in their bodies in order to influence buoyancy. This modification allows them to rise or sink slowly, or to simply float at any level.

Why do creatures make these journeys? There are various theories. As with humans, finding something to eat and someone to love are at the top of the list. Each time deepwater animals migrate, whether up or down, horizontal currents in the water column drag them to a different pasture. In shallow water there is more food. The chances to encounter the opposite sex are greater in shallower water, assuming populations mix more readily near the surface, where water movements are more vigorous than in the calmer deeper waters.

So why bother to live in deep water? Common wisdom suggests that a perpetually dark or dimly lit environment is a refuge from the visual predators that are dominant in the surface. This evasive strategy also works in reverse. Huge numbers of certain surface-dwelling copepods migrate downward below the photic zone (below 200 m) during autumn. These animals reduce the chance of being eaten during the winter, when food is scarce, and then return to shallow waters in the spring and summer to forage and reproduce when food is abundant. But remaining in one place too long can have serious consequences. Some surface predators, such as sperm whales, sea lions, the impressive sunfish, and certain seabirds, have adapted to follow their prey deep down. Some of these animals can reach 1200 m of depth on a breath

of air taken in the surface and feed in layers where deep-living prey hibernate. In other cases, diel migrants are consumed by nonvisual predators like gelatinous zooplankton and deep-sea fishes that wait and ambush vertical migrants as they descend. It's not always easy to hide and avoid predators in the twilight zone.

The pronounced and predictable mass movements of deep-sea creatures fundamentally alter the lives of other marine organisms and probably of mankind as well. One obvious consequence of vertical migration by billions of animals on a worldwide scale is that astonishing quantities of plants and animals are consumed in shallow waters. And so when the migrants return to the deep ocean, enormous amounts of food are exported downward, linking the surface ocean and the deep sea. Yet no one knows exactly how much material they remove, how deep they move it, or how much they recycle. Why? Scientists would reply that we haven't been able to assemble all the pieces of the puzzle. The oceans are vast habitats. The deep sea covers over two-thirds of this planet's surface and comprises about 99% of its volume. We simply don't have enough technology to survey everywhere. However, greater numbers of sophisticated instruments are placed in marine environments each year and record tremendous quantities of data. Although more information is generated about the ocean than ever before, many mysteries remain. Each time we find a new species, we realize there must be hundreds more yet to be discovered. •

Caulophryne jordani
Fanfin seadevil
_SIZE 25 cm, not including sensory filaments
DEPTH 700–3000 m

This monster never comes near the surface; rather, it lurks in hiding, motionless, at the limit of light's penetration threshold, deeper than 700 m. The Greek root of its name evokes a "toad embellished with stalks," in reference to the myriad sensitive filaments that allow it to detect the slightest displacement of water. The anglerfish in this photo is easily distinguished as a female; the males are much smaller. It is rarely captured or even observed.

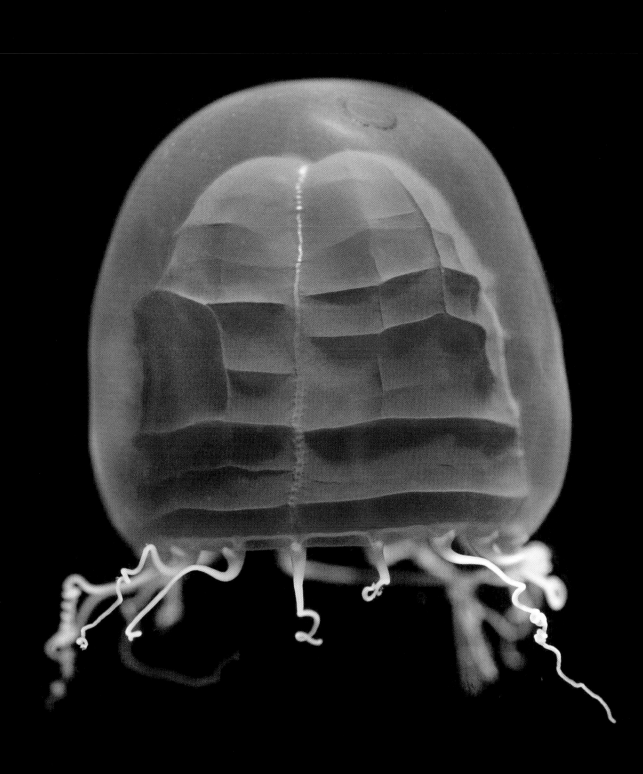

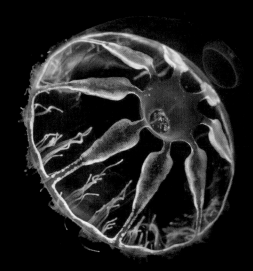

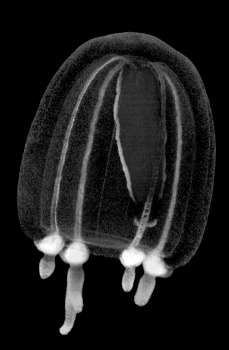

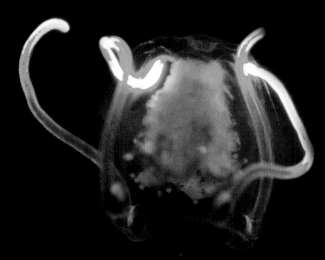

OPPOSITE
Pandea rubra
Red paper lantern medusa

_SIZE 15 cm
DEPTH 550–1200 m

The lantern medusa's ability to crumple and wrinkle its bright red umbrella, or to bend its form into right angles, is quite unusual for a gelatinous animal. When this very rare jelly was observed in the United States for the first time, scientists on board the oceanographic ship thought it was a new species. Its metamorphoses reminded them of the Japanese art of paper folding, so they called it the "origami jelly." Later they realized that this surprising creature had already been given a common name by biologist Dhugal Lindsay... in Japan!

LEFT
TOP *Botrynema brucei*
MIDDLE *Euphysa flammea*
BOTTOM Unidentified species

Each year, many unknown gelatinous creatures are found by scientists, but in order for the taxonomic description to give rise to the creation of a new species, the same organism must have been captured several times. Often our observations are singular events, and this is the case for the jelly shown at the bottom of this page; it will need to wait an indeterminate amount of time before finding its place in the classification tree of living organisms.

NEXT DOUBLE PAGE SPREAD
Leuckartiara sp.

_SIZE 10 cm long, including tentacles
DEPTH 0–200 m

Jellies can take on surprising colors and shapes. The yellow omelet-like mass includes the animal's sexual organs as well as its mouth. This one drags its long tentacles behind to capture prey.

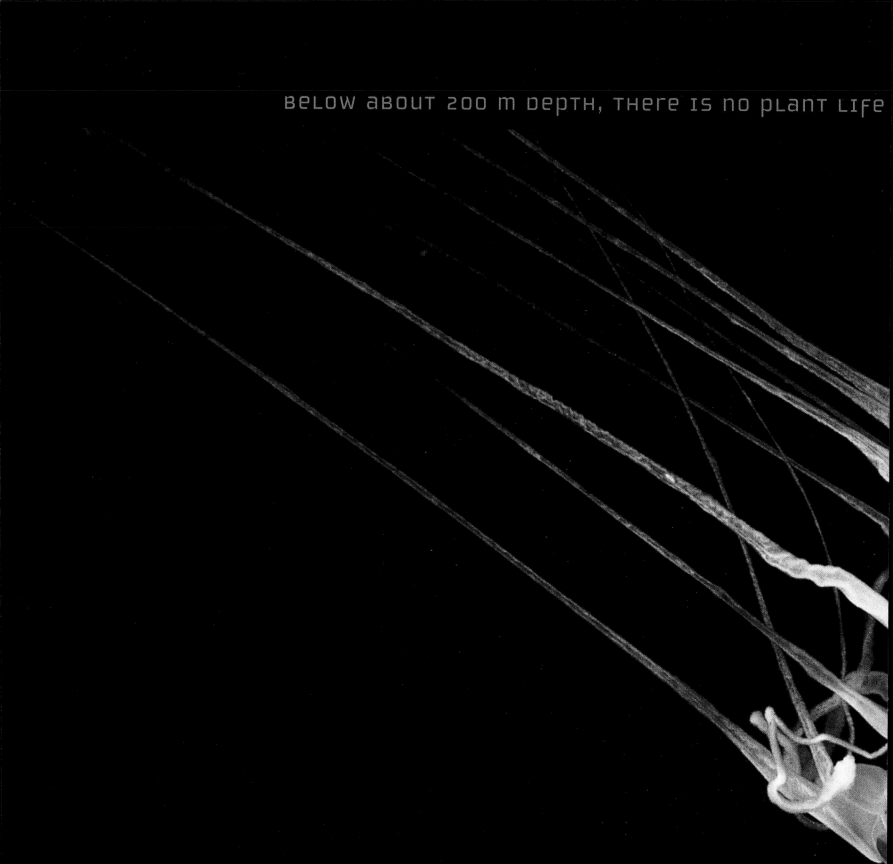

below about 200 m depth, there is no plant life

IN THE OCEANS, EVERYTHING IS EITHER ANIMAL OR MINERAL.

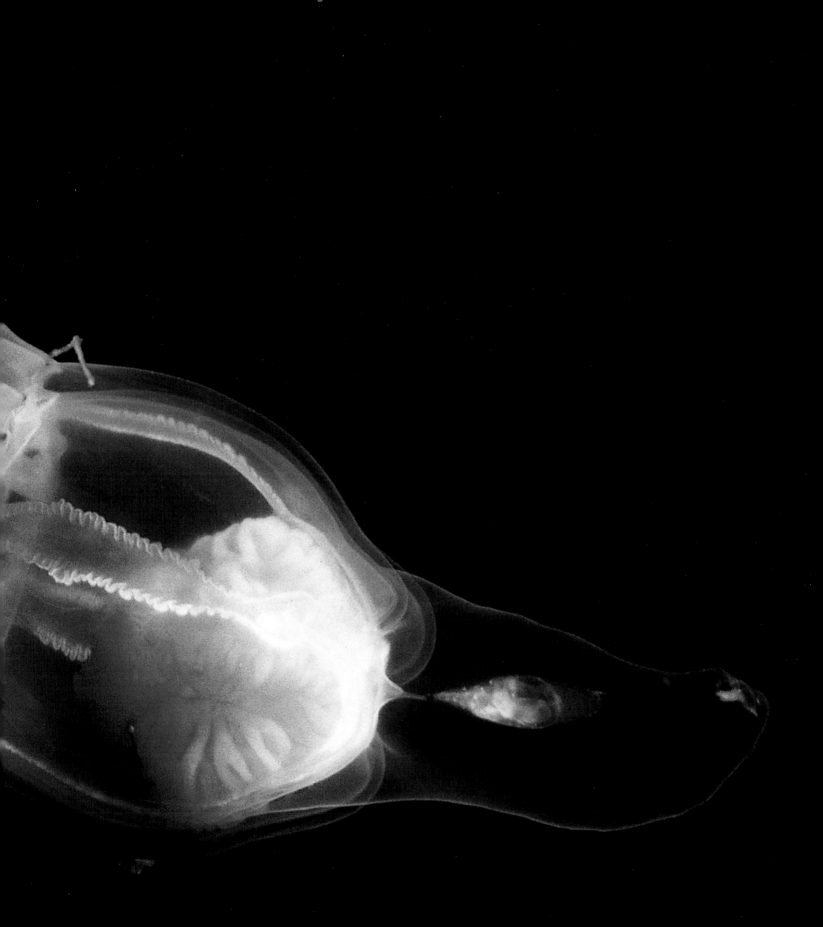

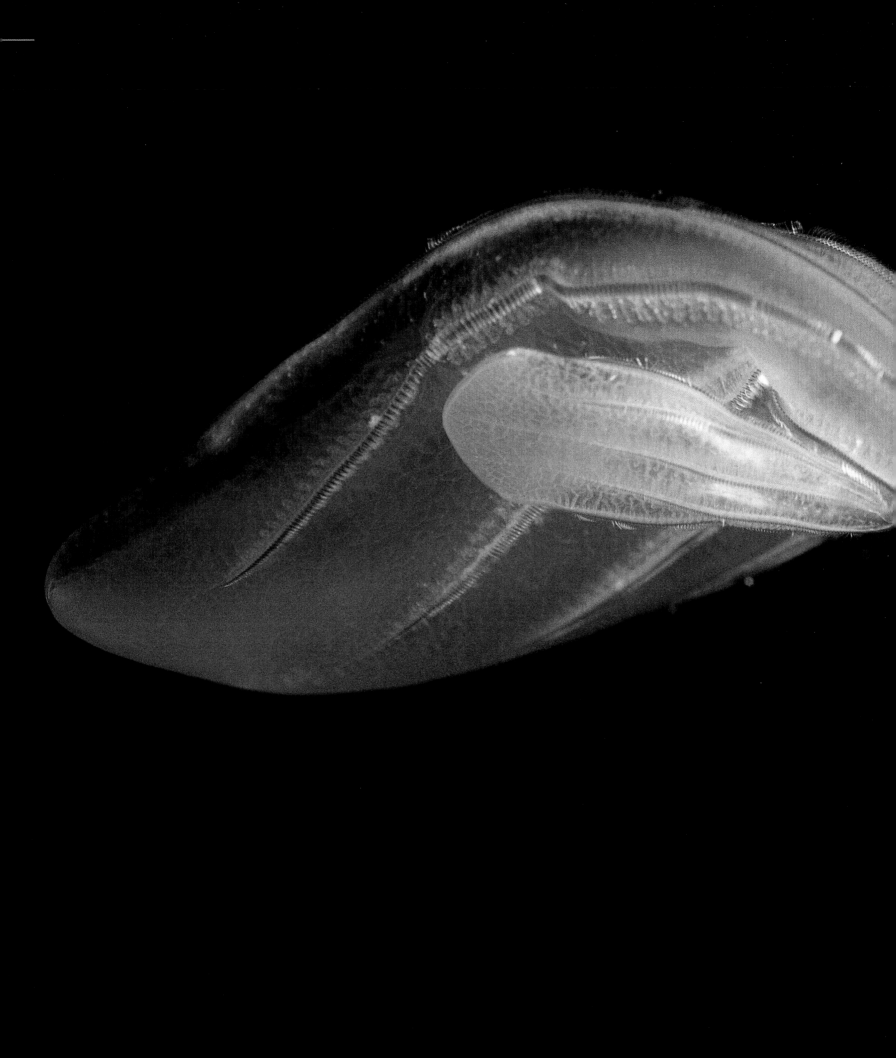

PRECEDING DOUBLE PAGE SPREAD, LEFT
Histioteuthis sp.

Jewel squid

_SIZE 25 cm
DEPTH 400–1200 m during the day,
0–400 m at night.

The jewel squid possesses particularly
sophisticated bioluminescent
photophores. These are equipped
with filters, reflectors, and
"eyelids," which allow it to regulate
the duration and intensity of its
luminous emissions, or even to turn
them off completely, based on
the animal's current depth and
the intensity of the ambient light.

PRECEDING DOUBLE PAGE SPREAD, RIGHT
Stereomastis sp.

_SIZE 10 cm
DEPTH 0–5000 m

This odd little creature with
its rumpled helmet is the larva of
a blind benthic crustacean, one
that frequently ends up in deep-sea
trawling nets. Perhaps it is the air
pocket created in its swollen
carapace that allows it to "fly away"
and reach the water column, where
it will develop in the company of
tens of billions of other larvae.
It will go through several mutations
until it reaches mature size, allowing
it to descend again to the great
depths where it will live as an adult.

LEFT
Beroe forskalii

_SIZE 10 cm
DEPTH 0–120 m

Ctenophores are composed of 98%
water, a little bit of muscle and
nerve, and some collagen to hold it
together. Though they lack a brain
and eyes, this does not hinder their
success as predators capable of
engulfing prey of their own size.

Stauroteuthis syrtensis

Glowing sucker octopus

_SIZE up to 50 cm
DEPTH 700–2500 m

This finned octopus often inflates
itself into the form of a tutu.
This bell shape posture may
correspond to a resting position,
but it may also indicate some
sort of predator defense, which
would not be surprising, considering
what an aggressive intrusion it must
seem to have some noisy, flashy
machine barging into the quiet of
one's retreat.

LIVING LIGHTS IN THE SEA

Dr Edith Widder
The Ocean Research & Conservation
Association, USA

The deep sea is often described as "a world of eternal darkness." That is a lie. While it is true that sunlight does not penetrate below 1000 m, that does not mean that it is a lightless world down there. In fact, there are lots of lights—billions and billions of them. These are animal lights and they serve many life-sustaining functions. There are lights for finding food, lights for attracting mates, and lights used for defense. All these lights are generated by a chemical process called bioluminescence. There are only a few creatures on land that can make light. Fireflies and glowworms are some of the best-known examples, but there are a handful of others such as some earthworms, click beetles, snails, centipedes, and fungi. These, however, are relatively rare and they do not play a significant role in the balance of nature. By contrast, in the oceans there are so many animals that make light that there are vast regions where as many as 80 to 90% of the animals collected in nets are bioluminescent. In the ocean bioluminescence is the rule rather than the exception.

The reason that there are so many animals in the oceans that make light has to do with the nature of the oceanic visual environment. Out away from shore, in the vast open ocean that forms the largest living space on our planet, there are no trees or bushes for animals to hide behind. But just as on land, prey need to hide from predators. Some animals hide by being transparent. Others hide by descending into the dark depths during the day and only ascending into food-rich surface waters under cover of darkness. And still others remain at depths below the penetration of sunlight and survive on food that sinks or swims into the depths. It is because so many animals in the ocean survive by hiding in darkness that the ability to make light is so prevalent.

For animals that spend their lives avoiding sunlight, a built-in headlight can be a very handy device. There are many fishes, shrimps, and squids that use headlights to search for prey and to signal to mates. Headlights may occur below the eye, behind the eye, or in front of the eye. Many headlights have a highly reflective surface that helps direct the light outward, much like a car's headlights. And as with some cars, some headlights can be rolled down and out of sight when they are not in use—a handy way of hiding that reflective surface and allowing the fish to better blend into the darkness. Most headlights in the ocean are blue, which is the color that travels furthest through seawater and the only color that most deep-sea animals can see. But there are some very interesting exceptions like dragonfish with red headlights that are invisible to most other animals but that the dragonfish can see and use like a sniper scope to sneak up on unsuspecting, unseeing prey. Dragonfishes also have blue headlights that they can use like high beams to see into the distance.

Other animals use glowing lures to attract prey. Much of the fecal matter and decaying foodstuffs that rain down from above are covered with glowing bacteria, which is why a glowing lure can easily be mistaken for dinner, when it instead signals an untimely death in a toothy jaw. Lures may dangle from fishing rods that poke out of the top of the head or out of the chin; they may even be found at the tip of a very long tail.

Light is also used for defense. Many animals that live in the twilight depths between 200 and 1000 m use a camouflage trick called counterillumination to obliterate their silhouettes with bioluminescence. At a distance individual belly lights called photophores blur into a light field that exactly matches the color and intensity of the dim filtered sunlight overhead. And if a cloud passes over the sun the fish, shark, squid, or shrimp either dims its belly lights or swims downward to maintain that perfect match. One of the fishes that uses this camouflage trick is called the benttooth bristlemouth (*Cyclothone acclinidens*) ; it is so common that it is believed to be the most abundant vertebrate on the planet. Imagine that! The most abundant animal with a backbone, and most people have never seen or heard of it.

Another common defensive trick is for the prey to release its bioluminescent chemicals into the face of a predator, just as a squid or an octopus releases an ink cloud. The light either blinds or distracts the predator, allowing the prey to flee into the darkness. Many jellies use this trick, as do shrimps and squids. There is even a fish, called the shining tube shoulder, that can shoot the equivalent of photon torpedoes out of a fleshy, backward-pointing pipe located just above its pectoral fin.

Still another use of light for defense is as a burglar alarm. Blaring horns and flashing lights on your car are meant to discourage a burglar because of the unwanted attention they attract; brilliant displays of bioluminescence serve the same purpose. When caught in the clutches of a predator, a prey's only hope of escape may be to attract the attention of a larger predator that will attack the attacker. Some of the most spectacular light shows in the ocean are burglar alarms. One of the best examples is the pinwheel display of the common deep-sea jellyfish, *Atolla*. It is a display that has to be seen to be believed; in the dark depths of the ocean it can attract the attention of a predator over 100 m away.

Bioluminescence occurs in all the world's oceans from surface to bottom and from coast to coast. Appreciating how animals use their lights is important to understanding this ecosystem that represents more than 99% of our biosphere. Various light-producing chemicals extracted from different animals have also proved enormously valuable in medical and genetic research. Living lights in the ocean are beautiful, mysterious, useful to humans, and absolutely essential to the animals that possess them. •

OPPOSITE

Unidentified anglerfish

_SIZE 15 cm
DEPTH 1000–4000 m

There are 160 species of anglerfish, though new specimens like the one here remind us that there still are many unknown creatures living in the ocean depths. All anglerfish have a lure, bearing millions of luminescent bacteria, which the fish flashes on and off while waving it to and fro like a fishing rod. Prey are attracted by the lure into the fish's formidable jaws, thinking they have found an organic tidbit glimmering in the darkness of the abyss.

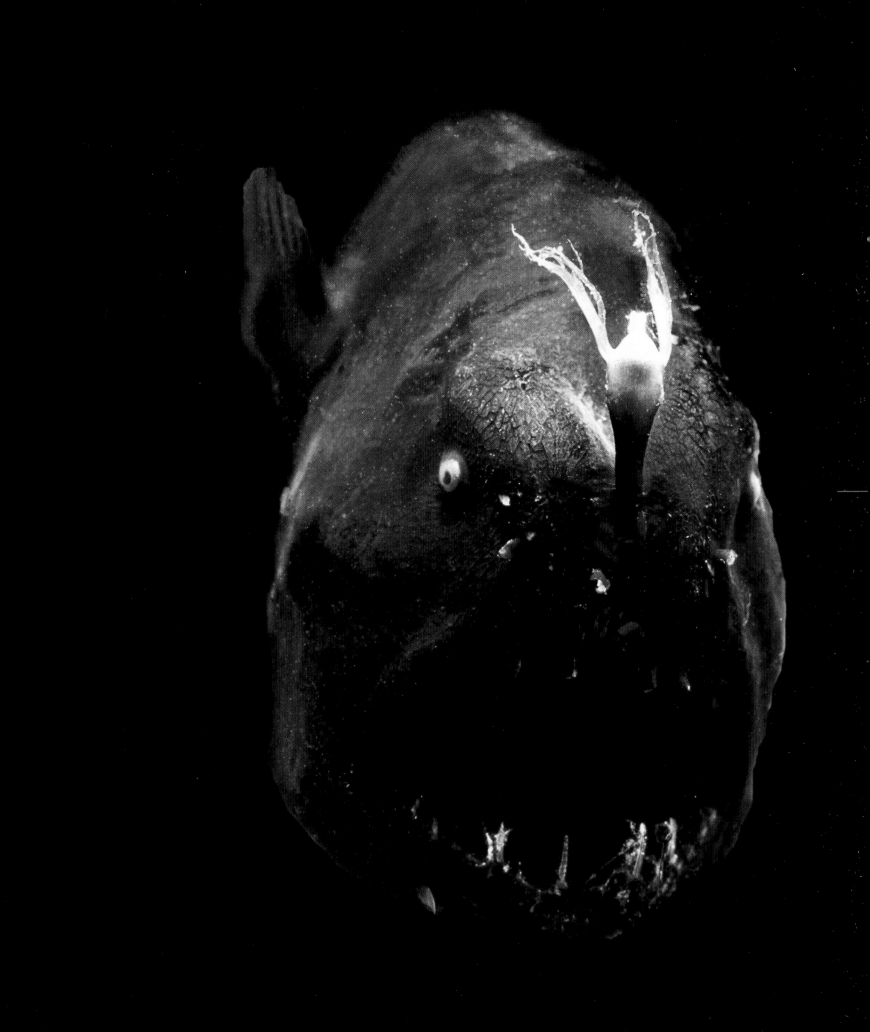

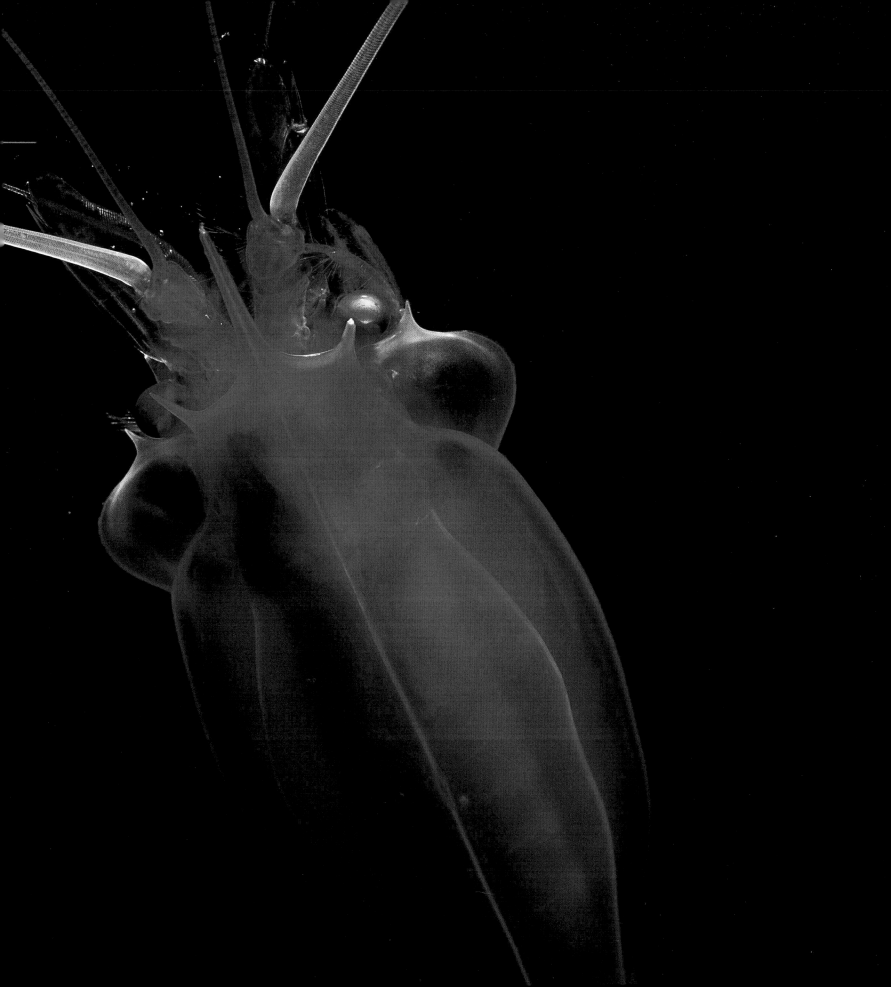

OPPOSITE
Gnathophausia zoea

_SIZE 10 cm
DEPTH 400–900 m

Out of water, the unique, surreal red
of this mysid is so intense it appears
almost to be lit from inside. Red is
the first wavelength to be absorbed
by the water, so down in the depths
where it lives, this relative of
the shrimp appears totally black.
If this camouflage fails and it is
disturbed by a predator, the mysid
will spit a secretion of blue
bioluminescent particles, surprising
and disorienting the enemy.

NEXT DOUBLE PAGE SPREAD, LEFT
Colobonema sericeum
Silky medusa

_SIZE 5 cm
DEPTH 500–1500 m

This reserved little jelly is easily
recognized by its white-tipped
tentacles that suddenly detach from
the body and light up when it is
attacked. By the time the predator
regains its senses, the silky medusa
has already vanished into the
darkness.

NEXT DOUBLE PAGE SPREAD, RIGHT
Tomopteris sp.

_SIZE from a few millimeters to 30 cm
DEPTH 0–4000 m

There are several species of
Tomopterid worms, which vary
in color from red to orange,
passing through violet to total
transparency. The animal's color is
not a characteristic of the species;
rather it depends on nutrition.
One feature that all the species have
in common is the ability to secrete
a yellowish, bioluminescent fluid
from glands at the tip of their
appendages. The bioluminescence
produced by the majority of animals
is blue, because the eyes of
deep-sea fauna are accustomed
to detecting it. The purpose of
this yellowish light, which remains
practically imperceptible to
the creature's neighbors, remains
a mystery.

"This hidden deep-sea environment dwarfs all other

earthly habitats combined. It is the ultimate reservoir

from which life everywhere draws sustenance."

Robert D. Ballard, 2000

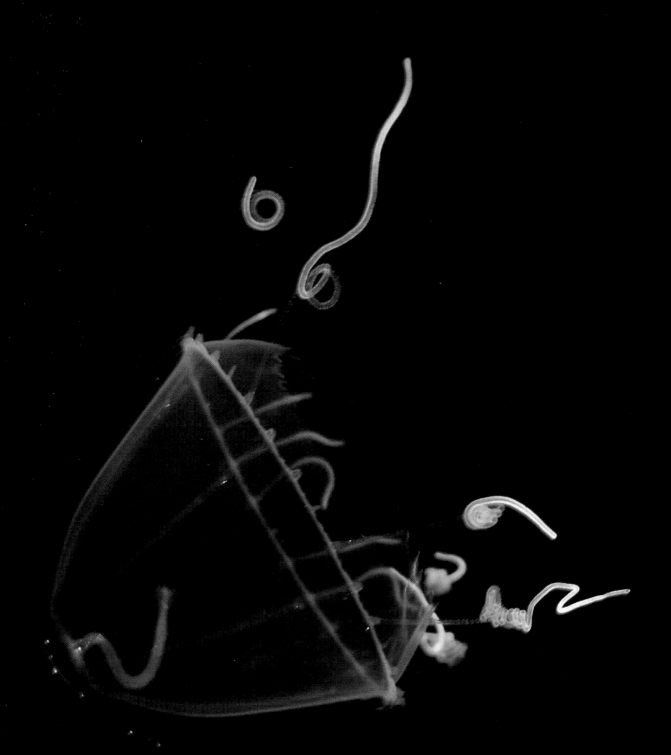

OWN LIGHT.
DOUBT THE MOST WIDELY USED MODE OF COMMUNICATION ON THE PLANET.

EVOLUTION CAUGHT RED-HANDED

The evolution of life is a huge puzzle, of which the *Stauroteuthis syrtensis* octopus is a very important piece. It is one of the rare octopuses that can produce a luminous emission from its suckers. While other octopuses use these to adhere to the substrate or to overcome resistant prey like bivalves, the glowing sucker octopus found another use for them as it adapted to the completely new conditions of life in the deep. It gets around the problem of food supply by luring prey, rather than chasing them in the awkward darkness of the deep.

This creature lives some dozens or hundreds of meters above the bottom, which is why it does not need adhesive suckers; instead, they have evolved into small individual lanterns. This strategy is particularly well adapted to its diet; it feeds almost exclusively on copepods, small and very abundant planktonic

crustaceans with excellent eyesight that are attracted to light sources like insects are to headlights. *Stauroteuthis syrtensis* secretes a mucous net that it holds between its arms with the help of its cirri—slender fingerlike appendages lining the underside of its mantle. When crustaceans approach the bioluminescent glow, they get trapped in the sticky net. Like whales, which feed on plankton, this creature also has to go for quantity in order to compensate for the small size of its prey. Although

bioluminescent photophores are widespread in other cephalopods like cuttlefishes and squids, it is extremely rare among octopuses.

Studying the origins and evolution of bioluminescence poses a real challenge, because there is no fossil record indicating how the organs have been modified or altered from their primary function to produce light. The glowing sucker octopus helps us understand how animals originally living near the surface successfully colonized the ocean depths, a critical phase in the evolution of deep-sea species. •

PRECEDING DOUBLE PAGE SPREAD
Thaumatichthys binghami
Wolftrap angler
_SIZE 9 cm
DEPTH 2432 m

This fish is as strange as it is rare. While the other anglerfish have bioluminescent lures in the form of a fishing rod or a barbel, the wolftrap angler hides its luminous organ inside its mouth, amid its long sharp teeth. Its upper jaw, which overhangs the lower by far, gives it the appearance of a rugby player who forgot to remove his mouth guard. Only some thirty specimens have been captured to date.

OPPOSITE BOTTOM, BELOW
AND NEXT DOUBLE PAGE SPREAD
Glowing sucker octopus
_SIZE up to 50 cm
DEPTH 700–2500 m

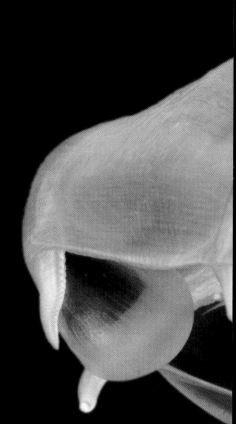

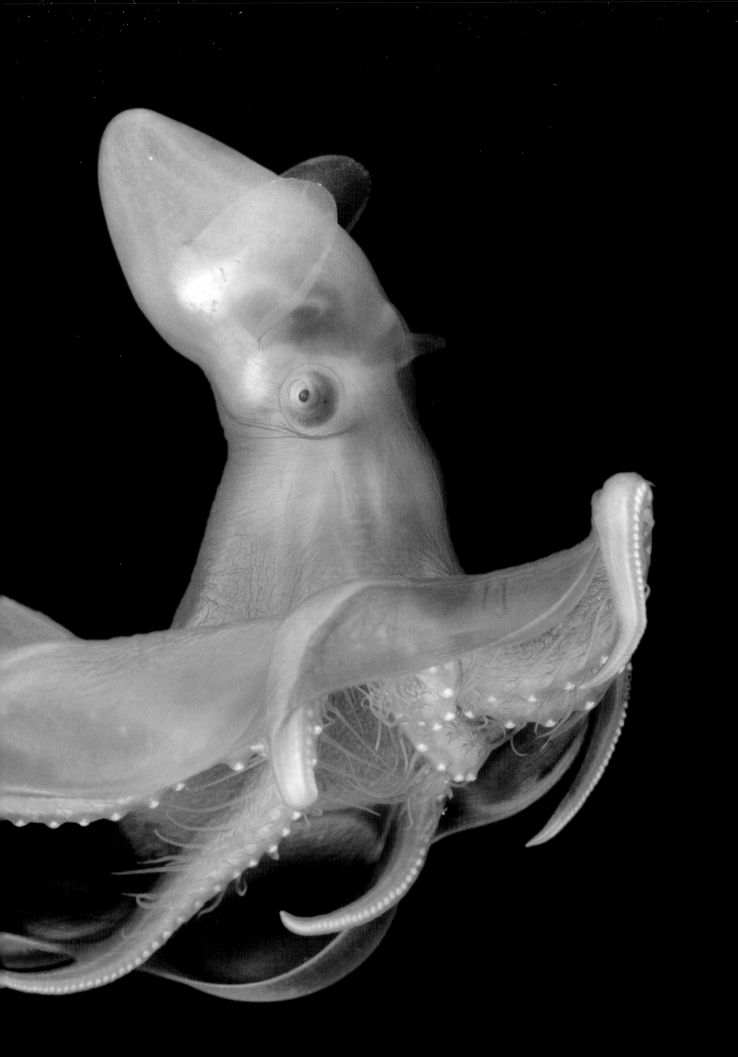

PRECEDING DOUBLE PAGE SPREAD

Stomias atriventer
Black-belly dragonfish

_SIZE up to 25 cm
DEPTH 100–1500 m

This dragonfish has all the
characteristics of deep-sea fishes
that are forced to save energy
in their food-poor environment:
its large eyes are adapted for seeing
in the twilight; its huge jaws are
equipped with sharp inward-curved
teeth; it has a fishing-rod-like
barbel with sensitive fibers
and a photophore for attracting
its prey; and to top it all off, it has
flanks covered with silvery scales
that reflect the available light,
causing it to blend perfectly with
its surroundings. These specialized
features make it one of the most
widespread fishes of the midwater.

OPPOSITE

Lampadioteuthis megaleia
Gem squid

_SIZE about 20 cm including tentacles

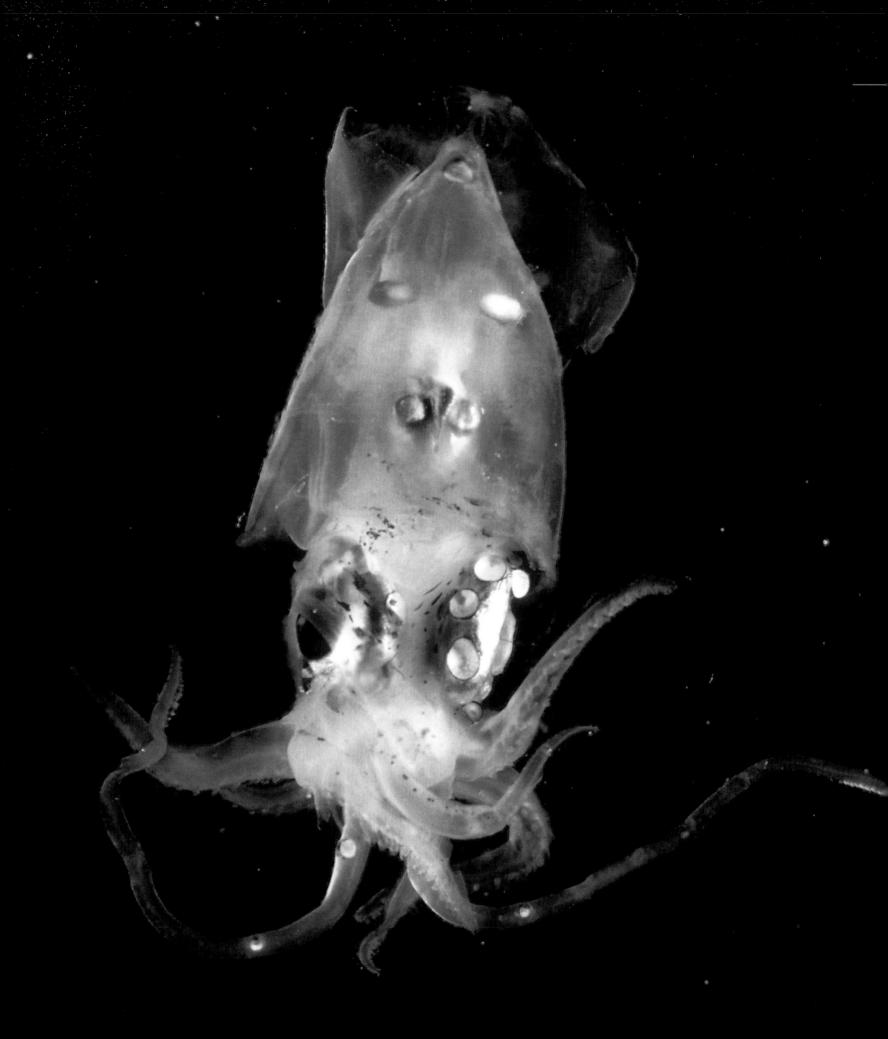

GeLaTINOUS BUT
VORACIOUS PReDaTORS

Dr. Laurence Madin
Woods Hole Oceanographic
Institution, USA

Ever since the invention of nets and trawls, man has been bringing up, among plentiful fish and cephalopod catches, various shapeless see-through gelatinous blobs that were hard to identify and often ignored by scientists. Sometimes, the blobs are animals such as medusae that can be made out and identified, but often it is impossible to even recognize body plans and to separate one animal from the next. This unattractive jello-like mass is the unfair land version of amazing and delicate creatures that can display their true beauty only in their natural watery environment. The importance and variety of these gelatinous organisms were greatly underestimated until the practice of scuba diving and underwater exploration vehicles were developed. What has been found in the past thirty years is that these transparent creatures are far more numerous, voracious, and widespread than first known, so much as to be now thought of as the dominant predators in much of the open ocean. Gelatinous invertebrates are so ubiquitous that they sometimes form a closed predation chain, with some individuals feeding mainly on other gelatinous species.

Who are these animals? Siphonophores, jellyfishes, medusae, ctenophores, snails, worms, and even some larval fishes. Most "jellies" have see-through tissues containing 95% or more seawater with just enough muscle, skin, and nerves to hold their bodies together and make them work. Fragile and transparent, without bones or teeth, brains or claws, some gelatinous

animals are among the most voracious predators in the sea. Many of these "organized waters," such as jellyfishes and siphonophores—the pelagic cousins of corals and anemones—have developed highly venomous weapons to catch and kill their prey while protecting their soft bodies. The graceful jellyfishes or medusae sometime trail toxic tentacles for tens of meters. Siphonophores are modular creatures, made up of specialized units: some just for locomotion, some for capture and digestion, some for defense, some for reproduction. They have a great range of sizes, from a few millimeters to over 40 m in length, Ctenophores, or "comb jellies," don't use poison darts but sticky glue cells to snag passing creatures: other gelatinous animals, small crustaceans, or even fish. Ctenophores move through the water by the action of rows of paddle-like comb-plates. The plates break light into prismatic colors, bathing the swimming comb jelly in ever-changing iridescence. In deeper water the ctenophore species are larger, with long tentacles or hugely expanded lobes of delicate tissue that stretch out to form silent traps in the dark water for unwary prey.

In surface waters, gelatinous animals all have a common feature: the transparency of their bodies, which makes them almost perfectly invisible to their predators. Or at least that's what we thought until we recently discovered that some regular predators of transparent animals like squids have broken the camouflage of their prey by using polarized vision—their eyes can detect changes in the polarization of light as it passes through the transparent bodies of prey. In the deep, below 500 m, many gelatinous animals have chosen on the contrary to darken their tissues to black or red, absorbing the blue-green bioluminescence that most animals create at those depths and particularly to cloak any bioluminescent sparks their prey might emit once in their stomachs.

In some areas, gelatinous animals are so abundant that they are capable of removing most of the food from the environment, being

OPPOSITE
Nanomia cara
Physonect siphonophore
_SIZE 2 m
DEPTH 400–1000 m

This siphonophore is equipped with gas-containing sacs—pneumatophores—for locomotion, venomous tentacles for capturing prey, and over eighty hungry stomachs. Periodically, these gracious and ethereal creatures become so prolific along the American east coast that they become the principal catch in nets, causing the collapse of local fisheries, since their gelatinous bodies have no commercial value.

direct and deadly competitors to other animals like fishes and cephalopods. Some jellies specialize in eating other jellies. The ctenophore *Beroe* engulfs other ctenophores whole, and deep-living narcomedusae snag and eat mainly other jellyfishes.

Many species can reproduce very quickly if the factors are right. They have low metabolic needs, which enables them to grow quickly on little intake, especially when much of their body is just seawater. We don't really know how long most can live, for they don't survive well in captivity, but small forms may have a life of weeks or months, and the large deep-sea species could well be decades in age. The deep-living colonial siphonophores might keep on growing and reproducing for centuries. All together, the jellies probably form the single most widespread group of animals in the oceans.

How can they be so abundant, make such a difference in the overall balance of the oceanic ecosystem, and yet have been ignored for so long? Large jellyfish in coastal waters have long been known, for they often wash up onto the shore. Scientific study of them began in the nineteenth century, when naturalists could take the fragile animals from the sea surface at places like Naples where currents brought them close to shore. But plankton studies soon became dependent on pulling nets through

the water, so that only the crustaceans and fishes were brought back, with jellies reduced to unrecognizable mush. Even while progress could be made in describing species, there was no way to observe them in their natural habitat until scuba diving was used for jelly research in the early 1970s. To get deeper in the ocean required other methods—manned submersibles and remotely operated vehicles (ROVs) . These tools extended our eyes and hands into the deep ocean and revealed how diverse and widespread the jellies are at every depth. Today perhaps two thousand species of jellies are recognized overall, with fifty or so new species being identified each year. The jellies live in all parts of the ocean, from shallow coastal waters to the deep abyss. Though their body plans seem simple, their designs have passed the test of time. Their evolution goes back over 500 million years, so they have been adapted to their oceanic lifestyle longer than most other animals; even the earliest Precambrian fossils look much like medusae. Every time we explore a new part of the ocean, we find new gelatinous species, sometimes tiny and obscure forms, but also deep-sea creatures nearly a meter across. Only now, after they have been here for millions of years, are we finally beginning to know these main inhabitants of our planet's largest ecosystem. •

Beroe sp.

_SIZE 15 cm
DEPTH 0 to at least 1000 m

These ctenophores are among the most active and voracious midwater predators. *Beroe* has no sticky or stinging tentacles for capturing prey, so it swallows them whole. It feeds exclusively on other jellies, sometimes of the same size, such as the comb jelly *Bolinopsis infundibulum*, seen here being devoured.

Pleurobrachia pileus
Sea gooseberry

_SIZE body, 1–2 cm; tentacles, 30 cm
DEPTH 0–750 m

The sea gooseberry is a principal constituent of plankton that produces up to a thousand eggs per day. It has two immense tentacles covered with adhesive cells, which it lets hang, like a flytrap, to snare the small crustaceans, eggs, and larvae that make up its menu. It then draws the tentacles to its mouth and starts the operation again.

Physalia physalis
Portuguese man-of-war, or bluebottle

_SIZE 30 cm

In contrast to the deep-sea jellies, this siphonophore lives at the surface. Like a curtain of death, its tentacles hang in a compact, threatening mass underneath the body. They inject the unfortunate prey with a potent toxin causing immediate paralysis that ends in a rapid death.

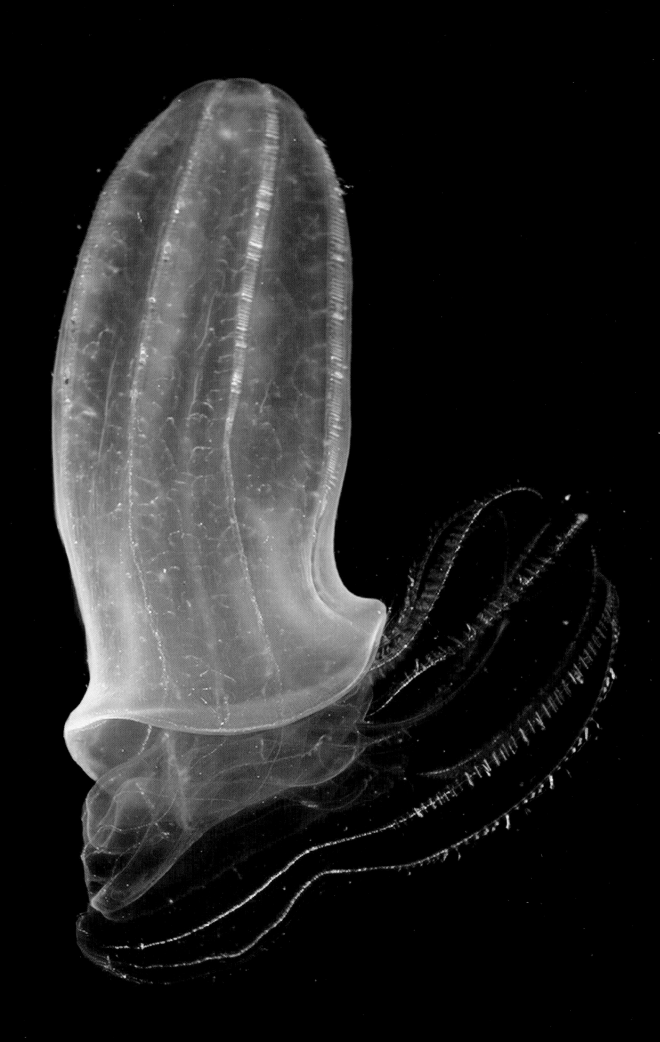

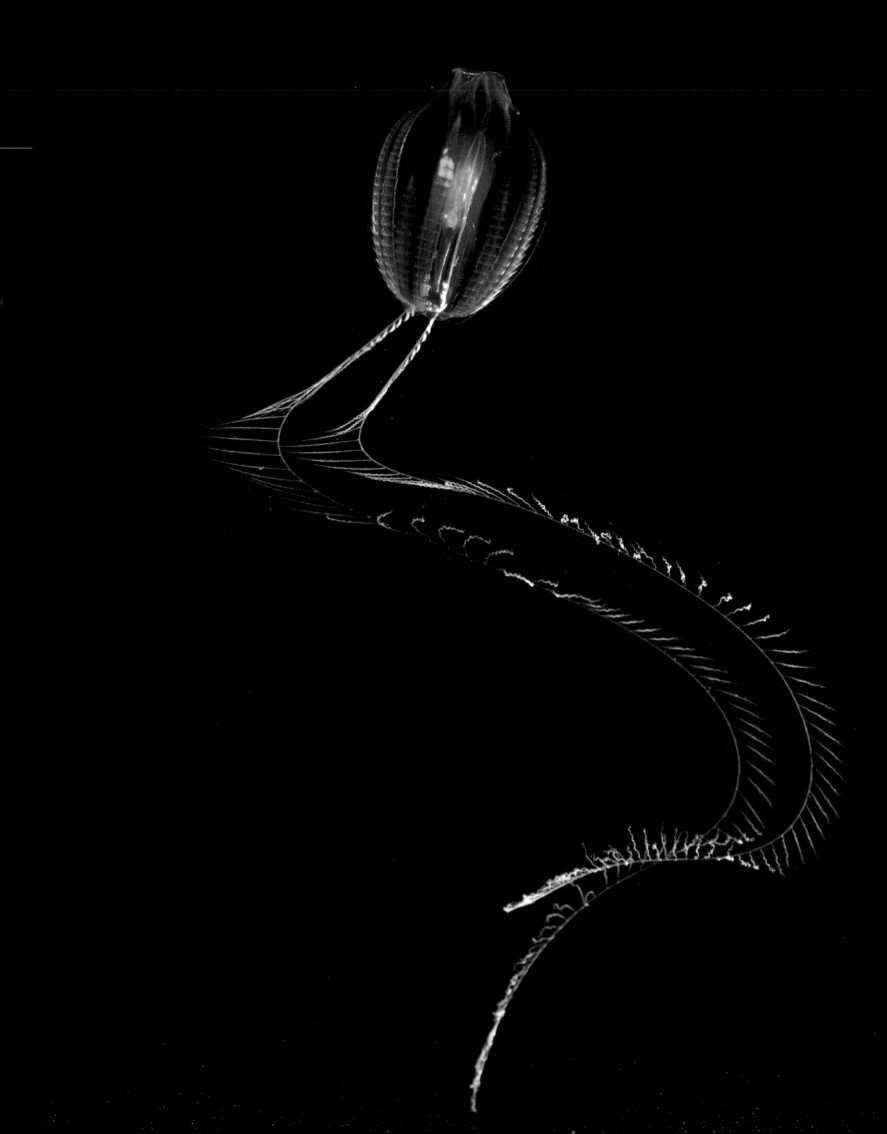

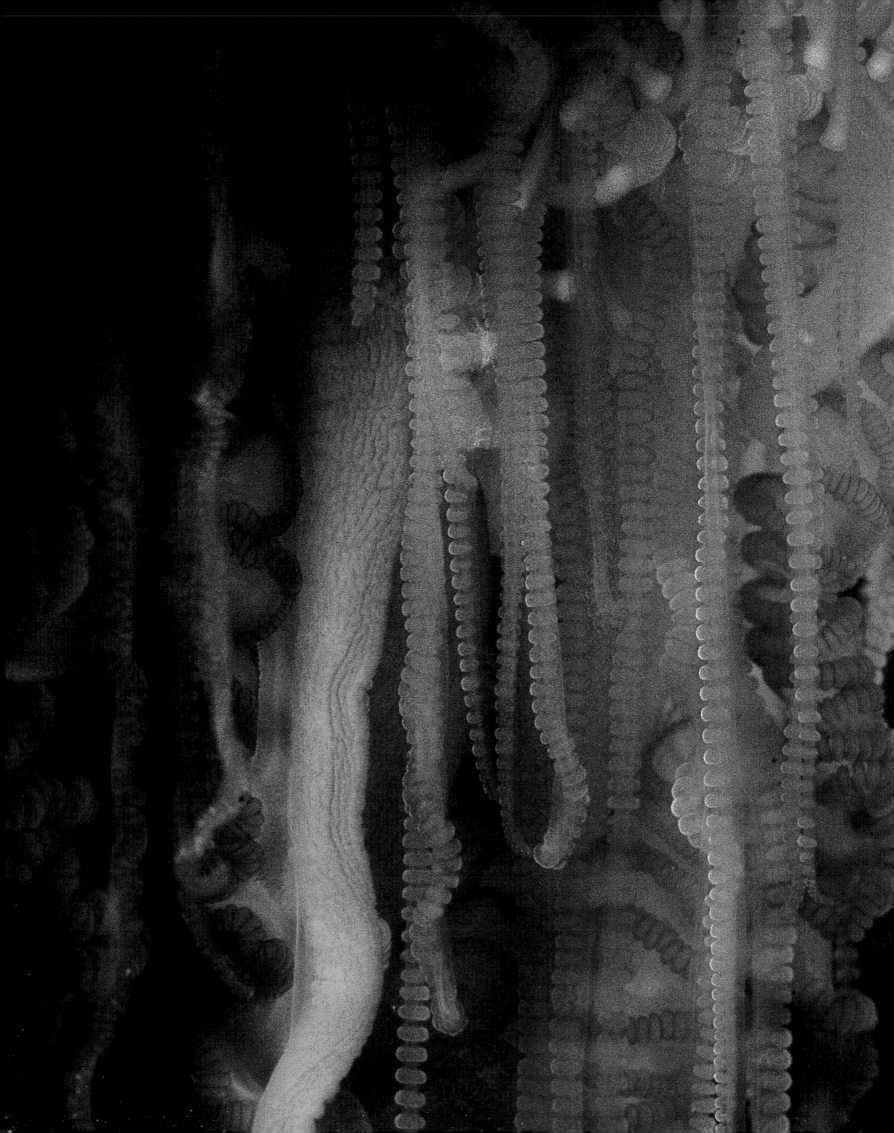

RIGHT

Unidentified species

_SIZE 2 cm
DEPTH 2000 m

The iridescent lights on this
ctenophore are not an effect of
bioluminescence, but the result
of the light reflecting off the
animal's cilia. These rows of tiny
oars beat, allowing it to move about
in the water. With neither eyes
nor brain, this creature hunts by
dragging its adhesive tentacles
behind; it can even accommodate
large prey owing to the impressive
size of its mouth.

NEXT DOUBLE PAGE SPREAD

Eukrohnia fowleri

Arrow worms

_SIZE up to 4.5 cm
DEPTH 700–1200 m

This small carnivorous animal,
which looks a bit like an orange ink
pen, is a member of one of the most
abundant groups of plankton
after the copepods. Their role in
the marine food web is not yet
well determined, although it is
probably significant in light of
their abundance.

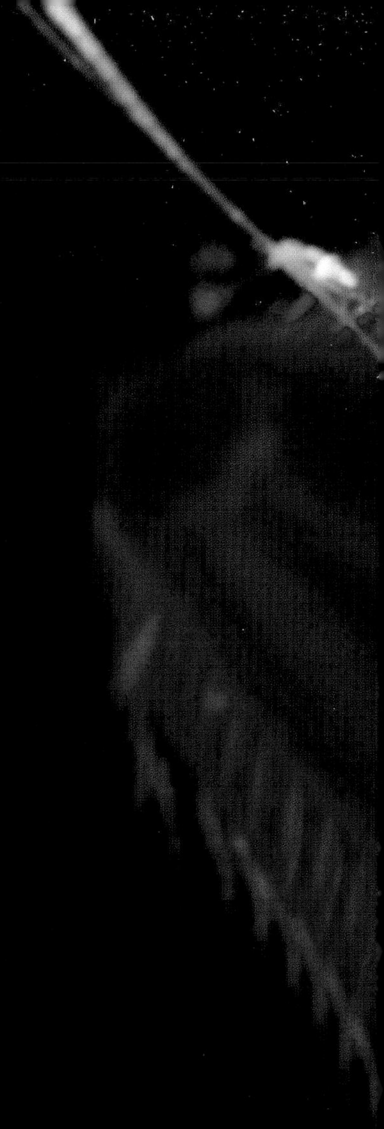

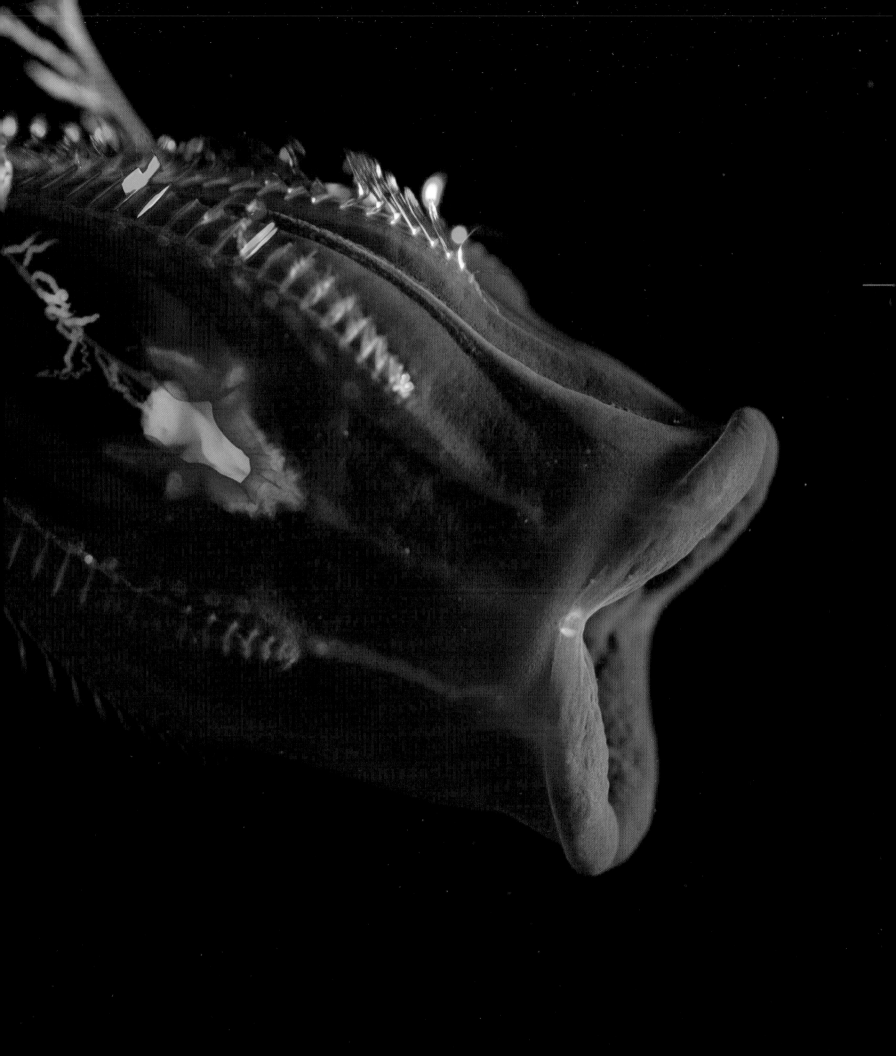

BODY IS EQUIVALENT TO A COW STANDING ON ONE'S THUMBNAIL.

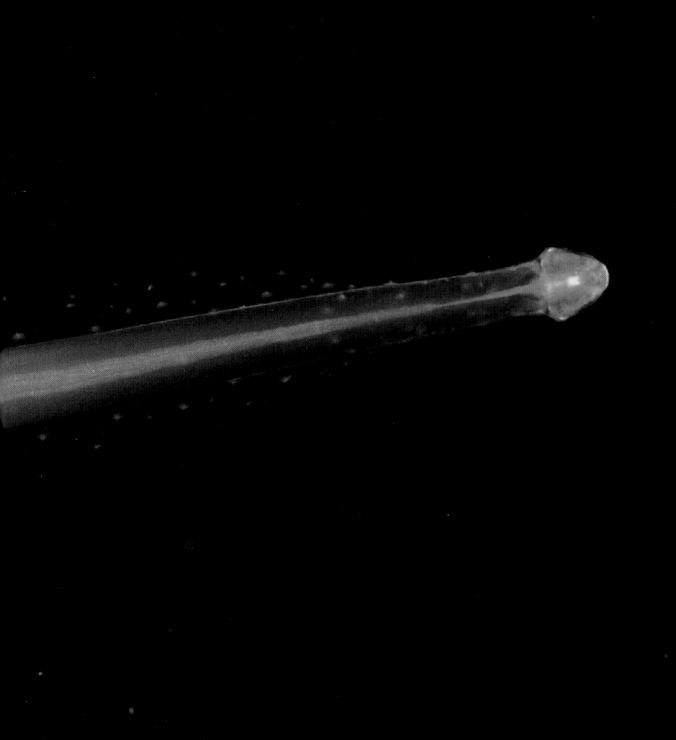

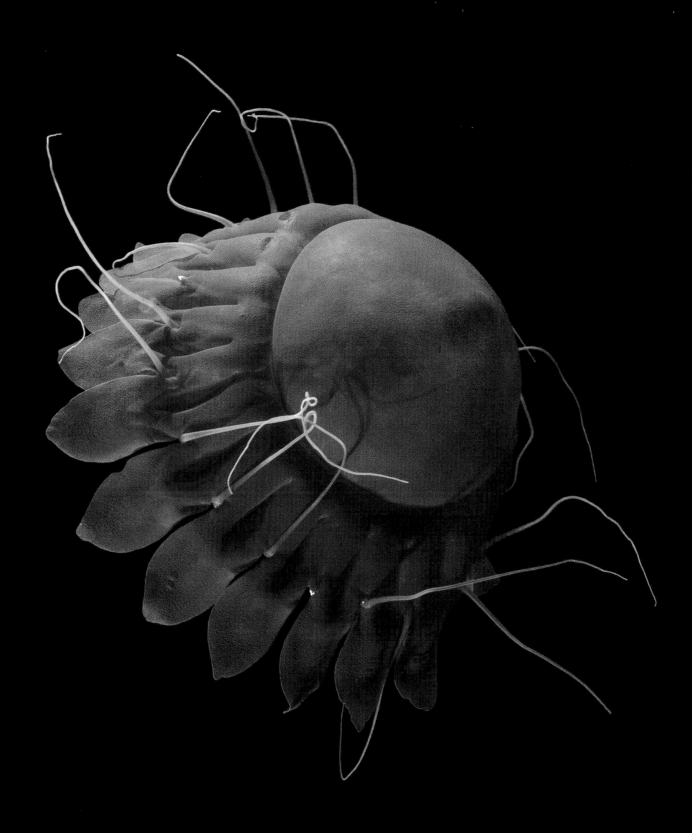

PRECEDING DOUBLE PAGE SPREAD, LEFT
Carinaria japonica
Sea elephant
SIZE up to 50 cm
DEPTH 0–100 m

This pelagic mollusk has conserved the features that mark it as a relative of land snails, but these are highly modified: its shell is reduced to a small triangular mass, seen to the lower left, and the foot that would have been for crawling on the ground has transformed into a fin, which it uses like a sail, always keeping it overhead. It is called the sea elephant because of the trunk hidden inside its mouth (to the upper right), used to swallow prey—mainly arrowworms and jellies.

PRECEDING DOUBLE PAGE SPREAD, RIGHT
Periphyllopsis braueri
SIZE 6 cm in diameter
DEPTH 600–1000 m

The red color of this bioluminescent jelly is a rather reliable indicator that it lives below 600–700 m depth. Above this threshold, most jellies are transparent, enabling them to dwell invisibly in this dangerous twilight zone.

OPPOSITE
Lampocteis cruentiventer
Bloody-belly comb jelly
SIZE 16 cm
DEPTH 700–1200 m

The stomach of this recently discovered ctenophore is always the deep red color of blood, although the rest of the animal's coloration can change. Red, which appears black in the water, allows it to mask the bioluminescent glow emitted by the prey it has swallowed and to digest it in peace without the risk of becoming someone else's meal.

"Phosphorescent glimmerings dance all around—large or small, steady or flashing—borne on creatures that seem to have captured the last light beams from now extinct stars, to light up their own existence in the domain of eternal night."

Albert I, Prince of Monaco, 1902

RIGHT
Apolemia sp.
_SIZE about 60 cm
Depth: 400–1000 m

OPPOSITE LEFT
Physophora hydrostatica
Hula skirt siphonophore
_SIZE 7 cm
DEPTH 700–1000 m

OPPOSITE RIGHT
Marrus orthocanna
_SIZE 40 cm
DEPTH 400–2200 m

Though they do not have powerful jaws, sharp teeth, or threatening fins, siphonophores are still veritable killing machines that number among the most voracious predators of the oceans. The curtain of stinging tentacles they deploy can attain a length of 40–50 meters in the largest of the known siphonophores, *Praya dubia*. Biologists Steven Haddock and Casey Dunn compare these superorganisms composed of specialized units to a train having a locomotive at the head for propulsion and various wagons trailing behind for reproduction, feeding, and defense. When the siphonophore moves, the "locomotive", buoyed by a gas-filled float, begins pulsing its swimming bells, seen at the top of the photos opposite.

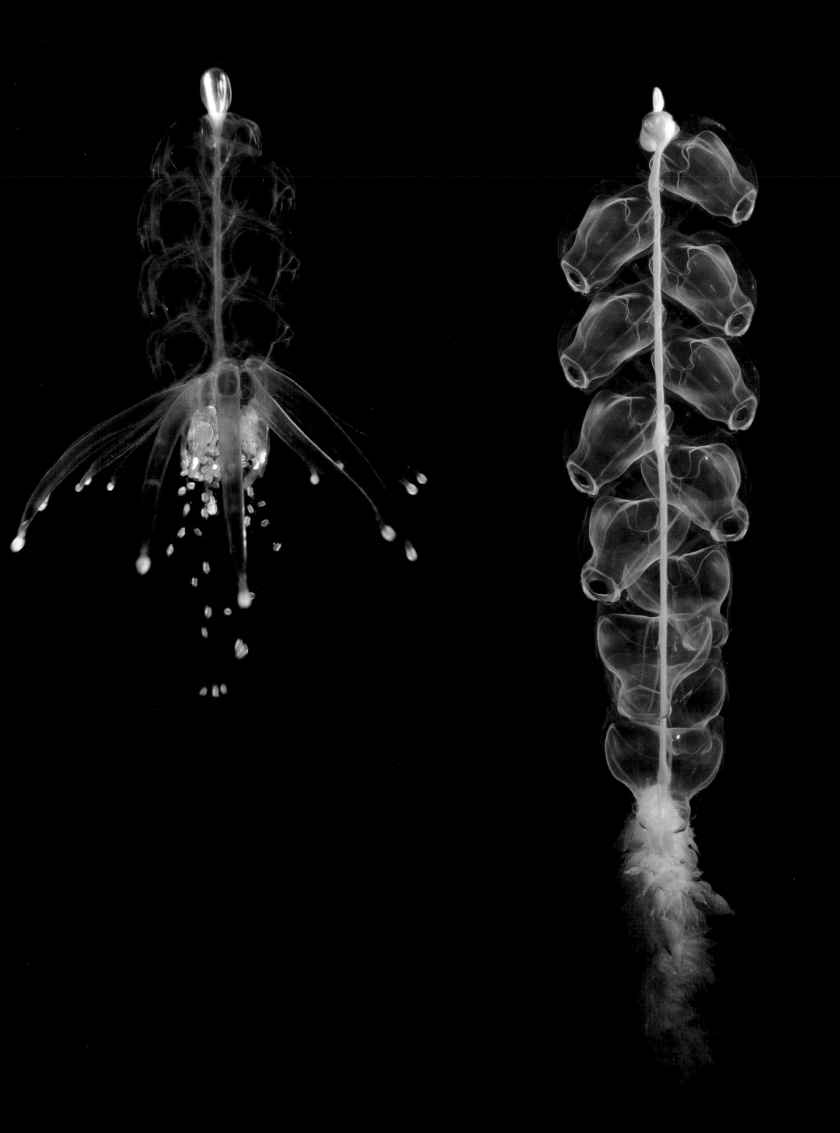

OPPOSITE
Phronima sedentaria
Pram bug
_SIZE 2 cm
DEPTH 200–1000 m

A pram bug wields its prey's remains as a shield, protecting itself with the hollowed-out barrel of a salp. The female lays her eggs and rears her offspring within the chamber. She often stands outside of the barrel and pushes it around like a pram carriage; hence its common name. Once the young reach maturity, they eat their own house before venturing into the midwater realm in search of other prey salps. According to popular legend, this was the species that served as the original inspiration for H. R. Giger's creature in the cult film *Alien* by Ridley Scott.

"Strange and beautiful things were brought to us

from time to time, which seemed to give us

a glimpse of some unfamiliar world."

Sir Charles Wyville Thomson, 1872

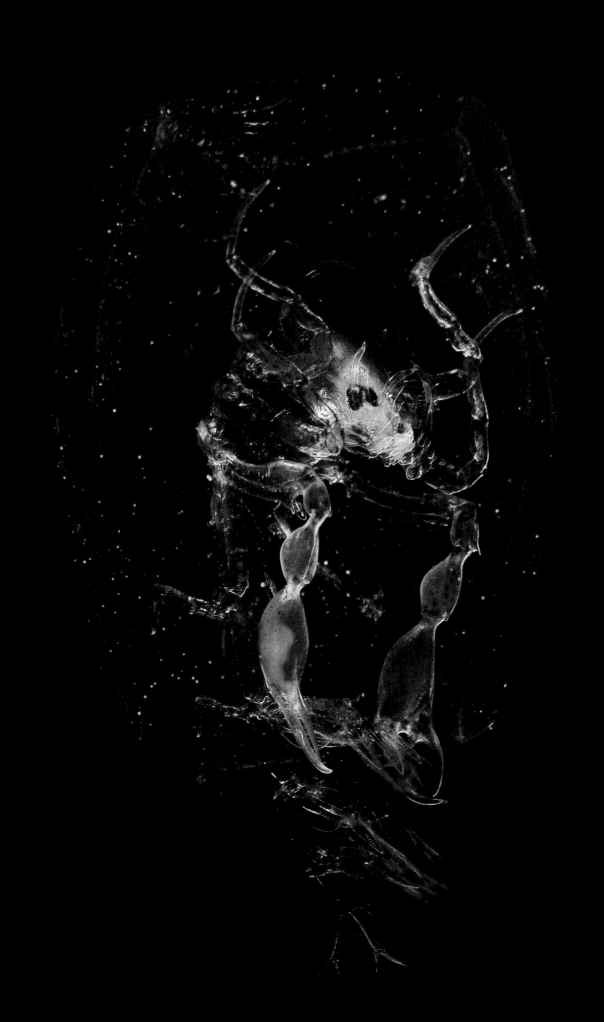

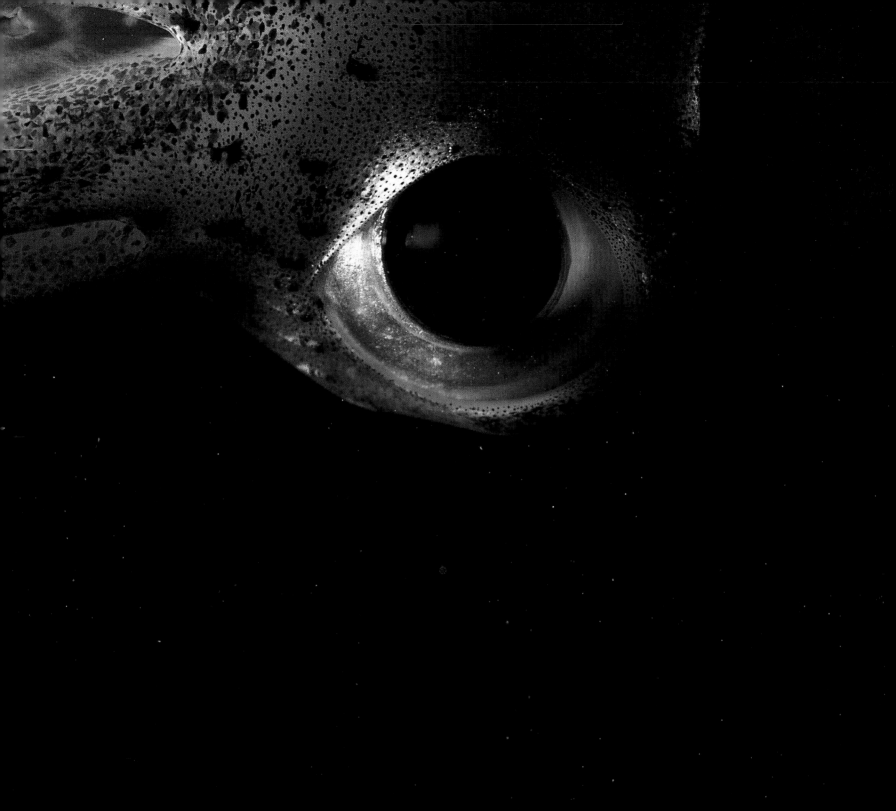

FROM MYTH TO REALITY:
MONSTERS OF THE DEEP

Dr. Clyde Roper
Smithsonian Institution, USA

Histioteuthis corona
Crowned jewel squid

_SIZE 40 cm including the tentacles
DEPTH 400–1200 m during the day,
0–400 m at night.

Symbolic of all the mysteries linked
to the quest for the giant squid,
this spellbinding, though somewhat
alarming, eye watches over us
in the night of the abyss. We know
from specimens washed ashore
that *Architeuthis* has enormous eyes
that can even reach the size of
a saucer.

The lore of the sea is shrouded in misty tales of
vicious monsters. Ancient mariners returned
from long sea voyages with hair-raising stories
and songs about many-headed creatures, huge
serpents, living islands, all monsters that attacked
ships and devoured hapless sailors. Occasionally
bits and pieces of rank tissue washed ashore
where old sailors recognized them as remnants
of the vivid monsters of their youth at sea.
The first written records of these mythical beasts
in natural history volumes of the 1550s were
illustrations and descriptions of sea monks,
mermen, and kraken; the Polish Bishop was
a particularly impressive sea monk. Since early
naturalists had never seen such beasts, and
no complete specimens existed, they took
the storied imaginations of sailors as the truth
and perpetuated the myths and legends
for centuries.

Two significant developments occurred
in the nineteenth century. The renowned Danish
zoologist, Professor Japetus Steenstrup, with
detective-like skill, revealed in 1854 that these
mythical monsters were nothing more than
squid. Not an overgrown freak of the common
calamari of coastal waters, this unique, huge
squid inhabited the depths of the open ocean;
Steenstrup named it *Architeuthis dux*, the king of
the chief squids. In a sense, this pivotal
discovery marks the beginning of the hunt for
the real giant squid. The second development
was a series of stunning discoveries that began
in 1873 in Logy Bay, Newfoundland, where
the first of a number of fresh, nearly complete
specimens was found; others appeared either
floating at the surface or stranded ashore
in the bays and coves of eastern Newfoundland.
These specimens permitted the first detailed
descriptions of this monstrous, though benign,
beast. About this time, a young author in Paris
learned about the discovery of a giant squid

off the Canary Islands by the French naval
corvette, *Alecton*. The captain had reported
sighting the beast at the surface and decided
to retrieve it for science. He fired upon it
with cannon and musket, but in the end decided
to abandon the battle, lest the monster endanger
his ship and crew. The story was not lost,
however, because the young author, Jules Verne,
stopped the presses on his latest novel and
incorporated the gripping saga of Captain Nemo's
encounter with a "squid of colossal size"
aboard the submarine *Nautilus*. So terrifying,
thrilling, and vivid were these scenes that
whenever we hear *20,000 Leagues under the Sea*,
we think "Giant squid!"

Through most of the twentieth century
a number of dead specimens were reported,
but the giant squid largely remained a mythical
creature. Principally because it never had been
seen alive, its habitat, distribution, biology,
and behavior remained unknown. Even though
marine biologists accepted the existence of the
giant squid, not much biology could be learned
in the absence of live specimens, other than
basic occurrence, morphology, and anatomy,
so they remained mysterious, legendary, almost
parenthetical members of the cephalopods.
Yes, we have learned that *Architeuthis* is
the largest invertebrate in the world, perhaps
up to 18 m in total length, 500, possibly 1000,
kg weight; it has the largest eyes of any animal,
the size of a human head! Over the years
we learned that it is a favored prey of the mighty
sperm whale and, in turn, it preys upon deep-sea
fishes like orange roughy and hoki, as well as
on other species of deep-sea squids. *Architeuthis*
inhabits depths from about 500 to 1000 m.
The sexes are distinct, with females much larger
than males at maturity. Today, over three
hundred specimens have been recorded from
around the world's oceans. While a number of

animals have been captured in deep-sea fishing trawls, most records are from strandings. This enigmatic situation exists because giant squid produce lightweight ammonium ions in their tissues, which enable them to maintain their buoyancy at their chosen depths without constantly swimming. When they die in the deep sea they float to the surface, where currents and winds drive some of them ashore. Our accumulated knowledge shows that *Architeuthis* is benign; this, however, has done little to hinder writers and filmmakers from perpetuating the horrible monster concept right into the twenty-first century.

The development of relatively dependable deep-sea cameras and lights by the mid-1990s enabled biologists and photographers to penetrate the inky blackness of the deep sea in search of a living, elusive giant squid. Cameras were lowered to the bottom on cables and left to cycle periodically whatever came to the bait, or they were suspended in the midwater on a buoyed line and allowed to drift through the depths to capture a glimpse of a giant squid. All attempts failed to produce images of *Architeuthis*. Perhaps as many as twenty to thirty searches that specifically targeted *Architeuthis* have been conducted, all to no avail.

A recent discovery adds to the excitement about the large creatures that live in the deep sea. A so-called "colossal" squid was captured by a long-line fishing vessel in Antarctic waters. This species has been known to specialists for nearly eighty years, based on very large specimens found in the stomachs of sperm whales, but it "went public" only with the recent capture of a huge specimen. The colossal squid

is not closely related to *Architeuthis*, and even a landlubber could distinguish the two species. Its body and head are longer and more robust than the giant squid's, but its arms and tentacles are relatively shorter. The geographical distributions of the giant squid and the colossal squid overlap only along the subantarctic convergence, at about 40°S.

Finally, in an event that excited ocean enthusiasts around the world, Japanese scientists announced that they had obtained the first photographs of a living giant squid in its natural habitat. Dr. T. Kubodera and his team lowered a still digital camera to 900 m depth, where studies indicated that sperm whales feed during their migration. Baited hooks mounted below the camera were attacked by a vigorous *Architeuthis* estimated to be 8.5 m in total length. The strobe lights exposed 550 images of the head and arm crown of the squid as it struggled to escape. When the camera system was retrieved, a broken tentacle remained attached to the hook, so identification of the specimen was confirmed both by morphology and DNA analysis. This spectacular event occurred after three years of search effort. It opens the door for further exploration where eventually we may be able to witness a sperm whale and a giant squid locked in a battle, one for its dinner, the other for its life.

After more than a century of study, *Architeuthis* has gained its rightful position as the largest permanent real denizen of the deep sea, a worthy representative of the most speciose and vast ecosystem on the ocean planet. We know that people need their monsters, and it is no surprise that the once-mythical squid of colossal size is the premier example. •

OPPOSITE
Computer-generated image
Mesonychoteuthis hamiltonii
Colossal squid

_SIZE 9 m
DEPTH larvae and juveniles 0–1000 m, adults 1000 to at least 2200 m

In contrast to the giant squid, which has suckers for capturing prey, the arms and tentacles of the colossal squid has hundreds of rotating hooks that bite into the flesh of its victims. These weapons make it a much more awe-inspiring predator, and more capable of larger prey, than the giant squid.

NEXT DOUBLE PAGE SPREAD
Computer-generated image
Architeuthis dux
Giant squid encounters the submersible *Johnson Sea Link*

_SIZE 18 m
DEPTH larvae and juveniles surface waters, adults 300 to at least 1000 m

Although legend describes the giant squid as an aggressive and relentless monster, the truth is otherwise; *Architeuthis* is apparently a timid creature that avoids contact and feeds on small animals. The myth has become so exaggerated that one finds it difficult to believe that, in an encounter with a sperm whale, *Architeuthis* is the one that gets attacked and not the other way around.

Are deep-sea Animals LIVING FOSSILS?

Dr Robert C. Vrijenhoek

Monterey Bay Aquarium Research
Institute (MBARI), USA

OPPOSITE

Careproctus longifilis
Threadfin snailfish

_SIZE 15 cm
DEPTH 1900–2997 m

Like a prehistoric tadpole, this fish
with a face perforated by large
sensory pores seems to confirm
the myth of the deep sea as a haven
for fossil creatures that have
remained unchanged since the dawn
of time. Despite its strange looks,
the threadfin snailfish is not
among the oldest sentinels of
our planet, as are the horseshoe
crab and the coelacanth, whose
fossil records date back more than
250 million years.

The deep sea is one of the most remote and unusual environments on our planet. As modern-day scientists and explorers, we are privileged to visit this alien world with our manned and robotic submarines and make discoveries that have shaken the foundations of our knowledge and beliefs. Earlier generations of scientists believed the deep sea was eternally dark, uniform, and essentially lifeless. This dark world would not support photosynthesis, the process that plants and many microorganisms use to obtain energy from the sun and make sugars from carbon dioxide. Instead, nineteenth-century scientists believed the ocean bottom was covered with *Urschleim*, the original slime that German philosopher and naturalist Ernst Haeckel proposed as the material from which all life arose. These beliefs collapsed with discoveries from the famous HMS *Challenger* oceanographic expedition (1872–76). Remarkable new species of animals were collected from the ocean depths with nets and dredges—predatory fish with grotesque heads and spiky teeth, and many new and unusual animals without backbones, some having monstrous shapes that suggested affinities with ancient mythological creatures. How unfamiliar and fantastic these animals seemed when compared with the animals of our surface world! Discoveries from the *Challenger* expedition stimulated imaginations and inspired myths about an eternally stable deep-sea world that is a home for prehistoric sea monsters. These myths were reinforced by the discovery in 1938 of *Latimeria*, the coelacanth fish from deep waters off southeast Africa. Coelacanths are lobe-finned fish that, like dinosaurs, left a diverse fossil record during the Mesozoic era (from 245 to 65 million years ago) but disappeared during mass extinctions at the close of the era. Discovery of this "living fossil" was as exciting to scientists as finding a living dinosaur. Was the deep sea home to other living fossils that survived the catastrophic extinction events that ravaged and revised biological diversity on the surface of our planet?

Though the deep sea is home to a remarkable variety of animals, research during the last twenty-five years has taught us that the ocean bottom is not eternally stable. Forces that create this instability can come both from the top down and from the bottom up. Climate change on the Earth's surface can lead to extinctions in the deep sea. Most deep-sea animals need oxygen to survive, and oxygen is brought to the deep by cold-water currents that sink from surface waters in the polar regions. During periods of global warming, surface waters become evenly warm and act as a cap on deeper water, the sinking process stops, and the deep water is trapped and deprived of oxygen, making animal life difficult. Such events may have led to mass extinctions of deep-sea fauna during the late Mesozoic (about 90 million years ago) and at the beginning of the Cenozoic (about 50 million years ago). Thus, many of the animal groups now found in the deep have evolved from ancestors that lived in shallower waters. The often stunning morphological features of deep-sea animals—huge teeth, big mouths, oversized eyes, bioluminescent organs—would then be a rather fast adaptation response to the dark, harsh environment of the abyss. Very few true living fossils have been recognized out of the thousands of deep-sea species identified during the past century. The vampire squid is one of them; it has changed relatively little in hundreds of millions of years. Its ability to live in oxygen-poor waters might be the reason this creature has survived for millions of years while so many other deep-sea animals have gone extinct.

The deep sea is not the perfectly stable environment we once believed because the seafloor itself is subject to constant change.

The great ocean basins have been reshaped many times by immense crustal plates that float on a sea of volcanic magma. Over time, these plates glide apart and collide with other plates, forming deep ocean trenches and creating gigantic mountain ranges. Spreading between these plates creates oceans like the Atlantic, which started forming 180 million years ago. Now, most of this spreading takes place along the midocean ridge system, a vast chain of mountains that encircles the globe much like the seams on a baseball. The ridge system is spotted with hydrothermal vents, hot springs created by the circulation of seawater through the heated crust. The discovery of deep-sea vents in 1977 near the Galápagos Islands astounded scientists. At depths of 2500 m, scientists found rich oases of life surrounding the vents. Large clams, mussels, snails, shrimps, crabs, fishes, and giant tube-dwelling worms flourish near the vents despite intense pressures, toxic gases, and heavy metals. Microorganisms use the energy stored in volcanic gases such as hydrogen sulfide and methane to fix carbon and make sugars. Perhaps the most enlightening product of these discoveries is our expanded view of the physical limits for life on this planet and a new realization that chemosynthetic life might also be found elsewhere in our solar system.

The discovery of such biological diversity at hydrothermal vents reignited the idea that the deep sea may be home to many ancient creatures. Surely, ocean ridges and hydrothermal vents have existed since the earliest days of our planet. The chemical environment at vents provides many of the raw materials that could have fostered the origin of life on Earth. Indeed, some of the most ancient microbial lineages are found at vents. Perhaps the geochemical energy from vents sustained many organisms during the cataclysmic events that led to mass extinctions on the surface of our planet. For example, the global mass extinction that led to the disappearance of the dinosaurs at the close of the Cretaceous era (65 million years ago) corresponded with the impact of a large comet that polluted the atmosphere with debris and probably reduced photosynthesis for many years. Would chemosynthesis protect hydrothermal communities from these surface events? Scientists hoped that they might find "living fossils" at vents, perhaps even a living trilobite, a group of arthropods that dominated the Paleozoic seas (543–248 million years ago). But these discoveries have not been forthcoming. Instead, the fossil record reveals that the vent fauna has been revised several times during the past 500 million years. Molecular biology suggests more recent origins of the animal groups that dominate present-day vents. The range of evolutionary ages seen among these vent animals is not too different from the ages seen in shallow marine faunas—perhaps a little older on average, but not much older. Communities of vent animals, so far from our eyes and daily activities, may be no better protected against environmental catastrophes than the plants and animals inhabiting the surface of our planet. Now, the rampant alteration and destruction of our planet's surface, the mining, the overfishing, and the pollution of our seas will likely impact deep-sea habitats as well. It is difficult to assess the threats that modern carbon dioxide emissions and global warming pose for the deep sea, but our studies of evolutionary history clearly teach us that this environment is no less susceptible to environmental disturbances than the Earth's surface. Expect to find no ancient monsters emerging from the deep sea, but look instead in a mirror and see an image of the modern creature that is perhaps the greatest single threat to biological diversity on this planet. •

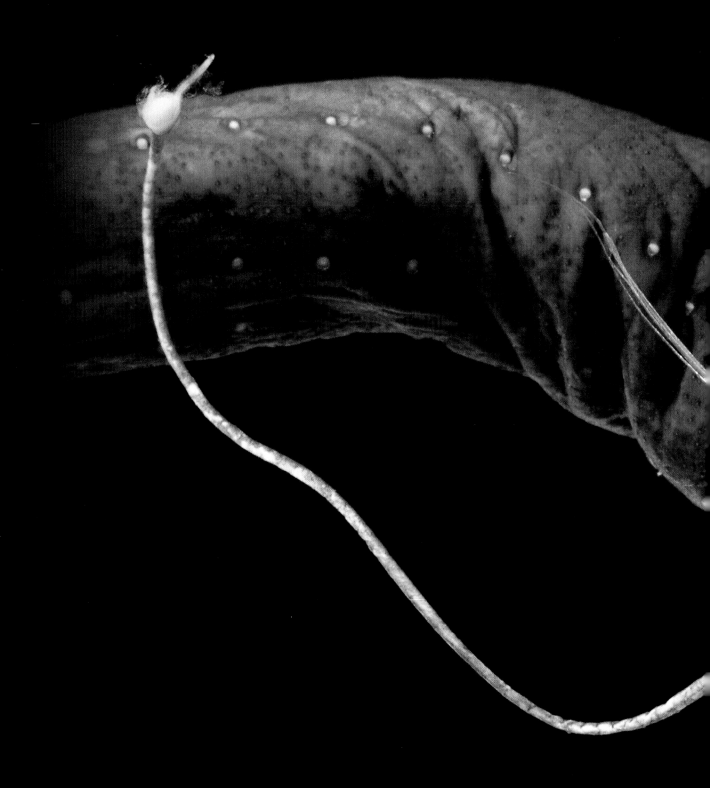

BUT THEY RARELY EXCEED A FEW CENTIMETERS IN SIZE.

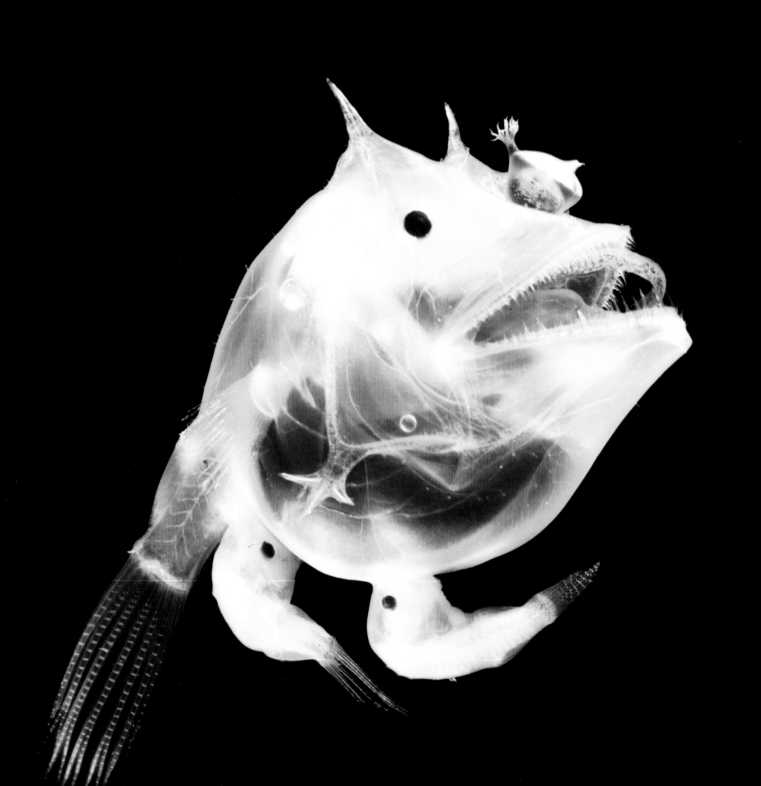

is in finding the rare females in the expanses of the deep. Once he is able to locate a mate, the parasitic male latches firmly onto the female with his hooked teeth. The only role evolution has reserved for him is that of a sperm pocket, with the sole aim in life to fertilize a female. The button-like protrusion between the fish's eyes is a bioluminescent lure.

"Strange creatures move about here and there—lilac-colored,

red, black—with organs that look monstrous to our eyes,

though they use them to walk, to swim, to crawl, to hold themselves steady,

to see, to sense, to fight... in a word: to live, in this habitat

where the conditions Nature has imposed upon life seem extraterrestrial."

Albert I, Prince of Monaco, 1902

described it in 1926 as "a very small but terrible octopus, black as night with ivory white jaws and blood red eyes." Neither of these two researchers had the chance to observe the vampire squid in its habitat,

that find shelter in the deep sea; its origins date back over 200 million years. It may even be the common ancestor of octopuses and squids, because it displays characteristics of both groups. It counts as an octopus with its eight arms and the fins on its head, like Dumbo octopuses, but also as a squid because

of its two long, retractable filaments that squids use for hunting. This remarkable cephalopod is so unusual that a separate order has been created for it, the Vampyromorpha. Another striking feature of this archaic creature is its ability to reside permanently within the oxygen minimum layer (OML). Available oxygen diminishes gradually with depth, being consumed by the numerous species populating the first oceanic kilometer, to the point of attaining a lower limit between 500 and 1000 meters, where oxygen levels may be no more than 5% of that available in surface air. Below 1000 meters,

oxygen is renewed by the masses of cold water sinking into the deep from the polar regions. The majority of cephalopods cannot survive where oxygen is less than 50% of surface air; they may wander into this layer for a few minutes, or even spend a few hours a day, but no others can remain there continuously.

How does *Vampyroteuthis infernalis* manage this physiological feat? By means of a respiratory blood pigment that can extract oxygen from the water very efficiently. This particular feature, along with the very slow metabolism of the vampire squid (the slowest of any cephalopod), allows it to go about its routine in an environment that is utterly hostile to other species. •

OPPOSITE BOTTOM AND BELOW
Vampyroteuthis infernalis
Vampire squid
_SIZE 40 cm
DEPTH 650 to at least 1500 m

When the vampire squid passes from juvenile to adult, it bears a double pair of fins on the head, like four incongruous ears. Little by little, the juvenile fins reduce until they disappear completely.
As a defensive measure, the vampire squid spits viscous, bioluminescent clouds from the ends of its arms, which can glow for up to ten minutes.

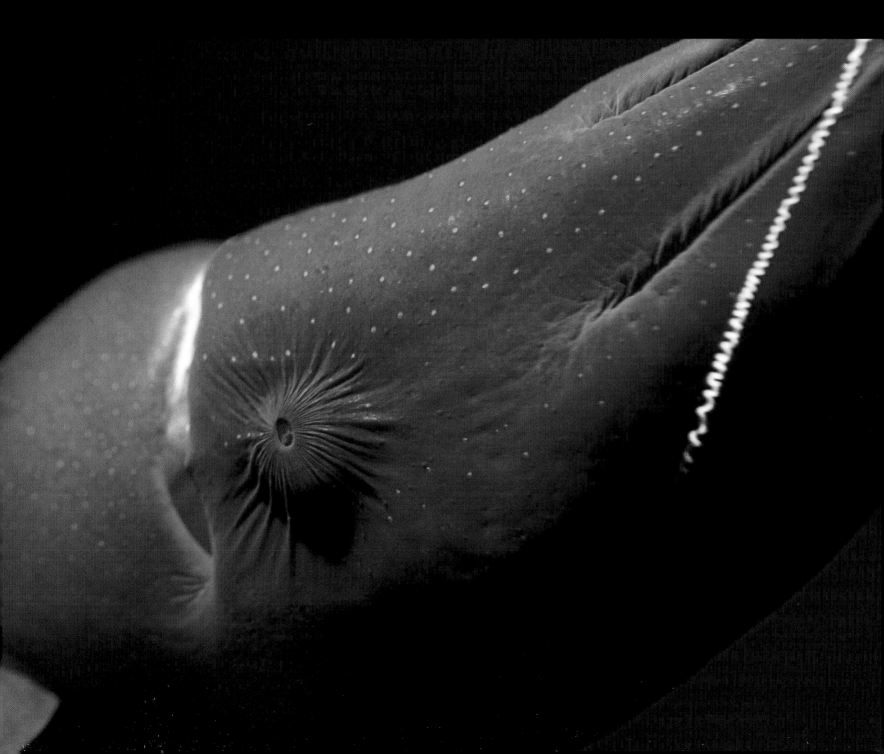

LEFT
Himantolophus paucifilosus
Football fish

_SIZE females up to 45 cm
DEPTH 1000–4000 m

With pearls dotting its body like
stitches, this football fish might
have been designed by Mary Shelley,
so strongly does it recall some of
Frankenstein monster's features.
Each of these dots is a sensory organ,
a neuromast, protruding through
the surface of the skin and allowing
the creature to detect the slightest
displacement of water. In surface-
living fish, neuromasts are embedded
within canals, otherwise they would
be too stimulated by the agitation
of waves and other vibrations.

TOP

Hymenocephalus italicus

Glasshead grenadier

Size 25 cm

DEPTH 100–2000 m

CENTER

Chiasmodon niger

Black swallower

_SIZE 10 cm

DEPTH 1500–4000 m

BOTTOM

Saccopharynx sp.

Gulper eel

_SIZE 2 m

DEPTH 2000–3000 m

Deep-sea fishes survive in a particularly harsh environment: it is cold, it is dark, and, to use Théodore Monod's expression, "it is hungry" down there. To endure this stark habitat, they have developed some rather remarkable features. The glasshead grenadier has enormous eyes that help it find its meals in the dead of night and a particularly long and streamlined tail that offers little water resistance. The gulper eel's oversized jaws make it look like some grotesque and ferocious Kermit frog. These two mandibles can open spectacularly wide in order to consume its prey whole. Very fortunately for the gulper eel, it has, like the black swallower, an expandable stomach that accommodates the size of its prey. This extreme adaptation recalls the python's capacity to swallow mammals that are much larger than the snake's girth.

OPPOSITE

Scopelogadus beanii

Squarenose helmetfish

_SIZE 12 cm

DEPTH 800–4000 m

The common name of this fish refers to the armored helmet worn by medieval knights. The unusually large holes over its face are nostrils, and the lacework of white strands covering its head is a network of sensory canals.

NEXT DOUBLE PAGE SPREAD

Stomias boa

Scaly dragonfish

_SIZE 32 cm

DEPTH 200–1500 m

The scaly dragonfish lures its prey with the bioluminescent organ hanging at the end of its chin barbel, and when they are close enough, it swings its jaws forward with lightning speed.

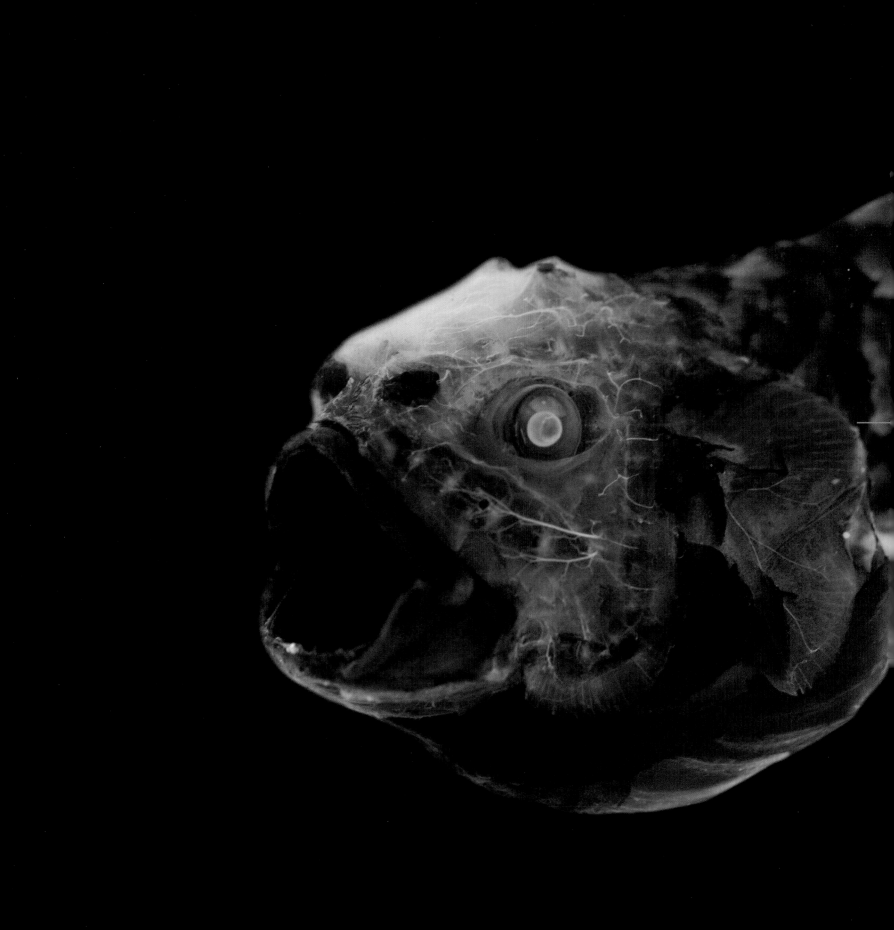

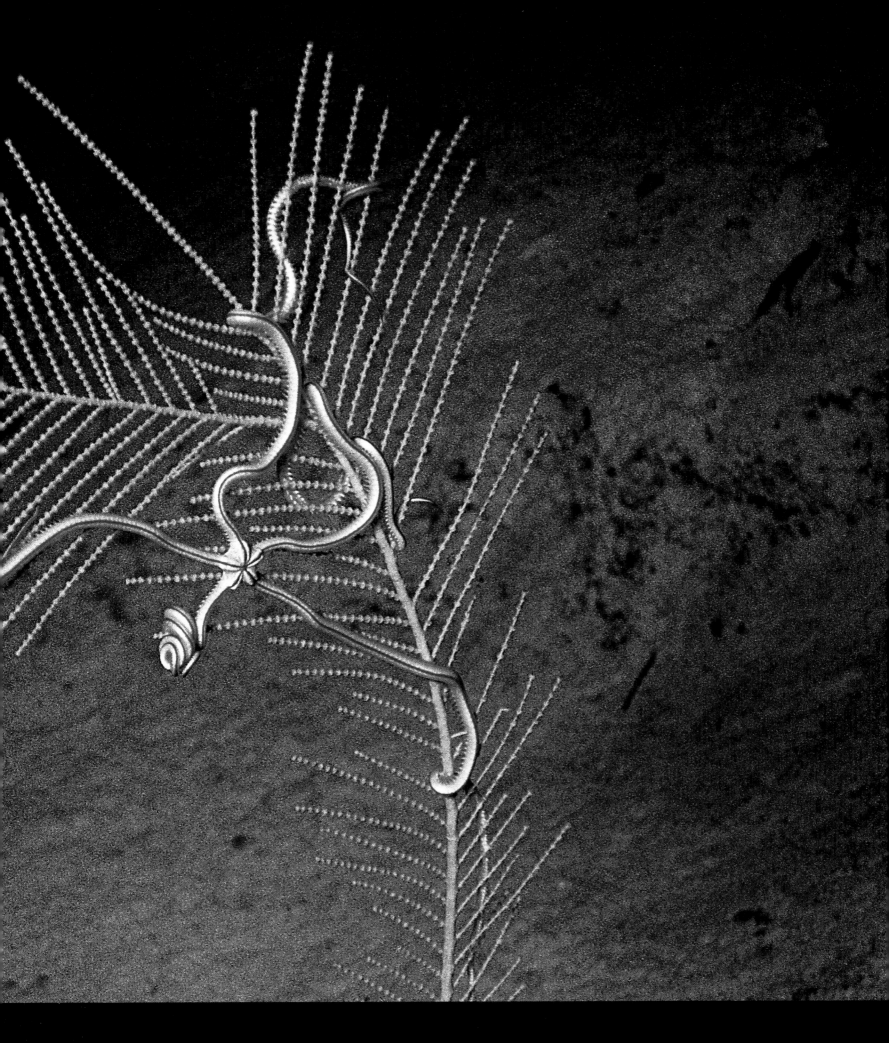

LIFE AT THE BOTTOM

We have long thought of the deep ocean floor as a vast muddy
expanse devoid of life. Today we know that the meager
resources coming from the surface are nonetheless
sufficient to feed a highly diverse benthic fauna, perhaps
as many as tens of millions of species. The surprising
discoveries of gigantic deepwater coral reefs and of densely
populated ecosystems, like seamounts and hydrothermal
vents, have completely altered our perceptions of the ocean

floor. These findings have revealed that the deep seafloor

is a very diverse habitat. In fact, chemosynthetic animal

communities, which draw their energy from toxic fluids,

amount to a biomass eight thousand times as great as that

found throughout the abyssal plains.

The following chapters invite us to satisfy our curiosity about

the deep seafloor's newly discovered ecosystems.

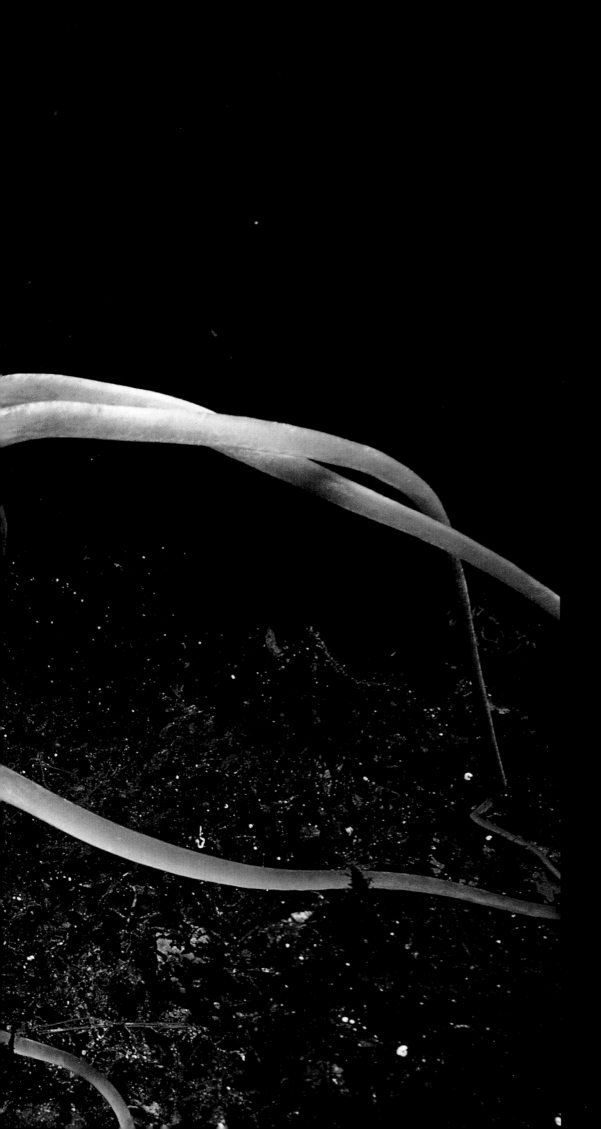

Asteroschema sp.

_SIZE 30 cm arm span
DEPTH 800 m

Brittle stars sometimes coil their serpentine arms around another animal, such as seen here with a gorgonian (*Calligorgia* sp.), in order to nab a meal from the current. When attacked, they will freely abandon an arm to a predator, through the phenomenon of autotomy, well known among lizards. The missing appendage then regenerates slowly.

PRECEDING DOUBLE PAGE SPREAD

Gorgonocephalus caputmedusae
Gorgon's head, or basket star

_SIZE central disk 6.5 cm
DEPTH 50 to at least 300 m

Shaped like an impenetrable ball of arms and branches, which it deploys when feeding, the gorgon's head is a relative of the starfish and the brittle star. It prefers dwelling in a place exposed to currents, unfurling its limbs very slowly, with infinite grace

LEFT

Periphylla periphylla
Helmet jelly

_SIZE up to 1 m
DEPTH 0–7000 m

This spectacularly enormous jelly unexpectedly proliferates in certain Norwegian fjords, where it is up to a thousand times as abundant as in the oceans. In the last twenty years, helmet jellies have become the principal predators in these partially isolated ecosystems, devouring all available resources, making it a dangerous competitor for the rest of the fauna.

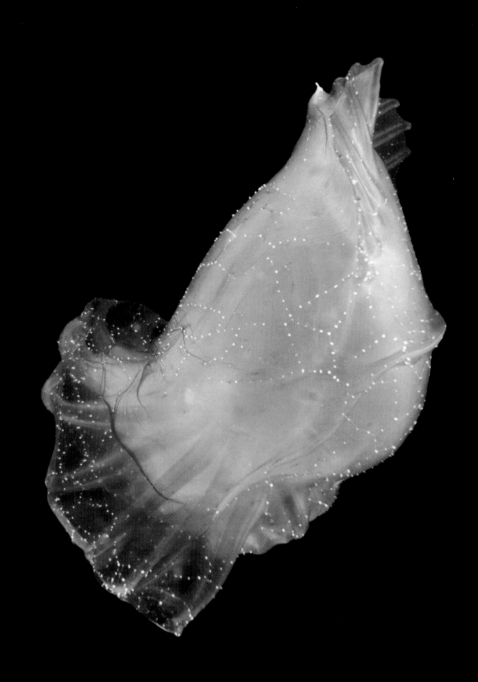

Archaeopneustes hystrix

_SIZE 20 cm
DEPTH 300–650 m

Like an army of tiny soldiers,
this congregation of urchins crosses
the abyssal plain, grazing on
particles of organic detritus in
the sediment. This species almost
always lives in groups of up to
sixty individuals, likely held together
by their contacting spines.

Enypniastes eximia
Enypniastes eximia

Deep-sea Spanish dancer
_SIZE up to 35 cm
DEPTH 500–5000 m

This sea cucumber can leave
the substrate for the water column,
sometimes up to several dozen
meters away from the bottom.
The deep-sea Spanish dancer uses
an unusual defense mechanism:
when attacked, its grainy skin lights
up and detaches, sticking to the
aggressor. Finding its face plastered
with a sticky, bioluminescent mask
that it cannot shake off, the would-be
predator becomes the vulnerable prey.

Alicia mirabilis

_SIZE 40 cm
DEPTH 10–300 m

Only experts were able to find a sea
anemone in this strange shape
laden with gelatinous protrusions.
Anemones sometimes move about
with irregular flip-flop movements,
in which case they draw their
urticating filaments inside the foot;
this causes them to look almost
like sea cucumbers. This animal was
photographed by a submersible pilot
in the Cayman Islands.

LIFE AT THE BOTTOM

151

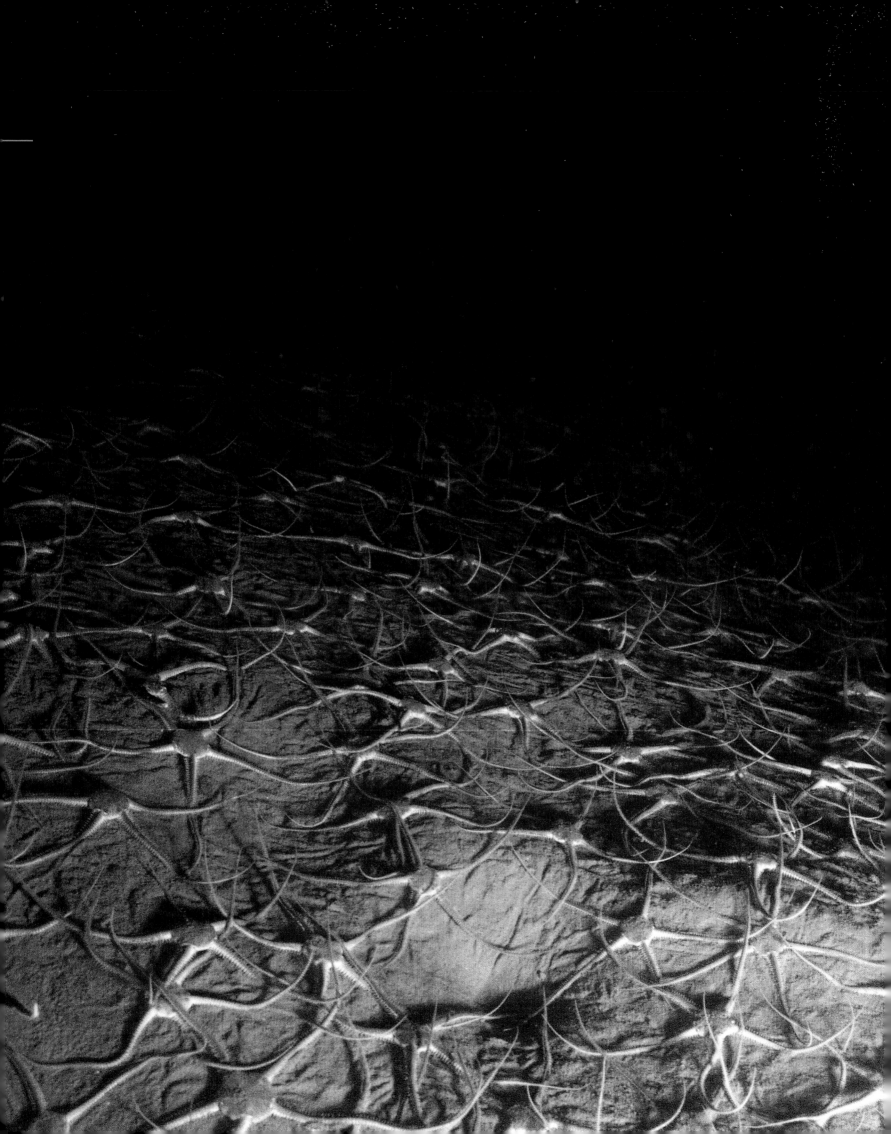

OPPOSITE

Ophiura sp.

_SIZE 25 cm
DEPTH 900 m

Colonies of these brittle stars
can number hundreds of individuals
per square meter, contradicting
the idea that the deep seafloor is a
lifeless desert. Why do these cousins
of the starfish pile on top of one
another like this? One possibility is
that these gigantic reunions occur
in places particularly well-exposed
to a flow of food particles; another
is that the creatures are responding
to some sexual calendar.

Dr. Craig M. Young
Oregon Institute of
Marine Biology, USA

THe Deep Seafloor:
A Desert Devoid Of Life?

Once thought to be a flat and lifeless desert,
the deep seafloor is now known to have more
topographical relief than the Himalayas and
a diversity of animal life that may exceed that of
the Amazon Rain Forest and the Great Barrier
Reef combined. Whereas the average elevation of
land is about 0.75 km, the average depth of the
sea is on the order of 3.2 km, making the deep
ocean the largest habitat on the planet; the deep
ocean floor is the largest surface on Earth
that supports animal life. Indeed, it has been
calculated that if all of the dry land on Earth
were pushed into the sea with bulldozers, it
would fill only about one-twenty-third of the
ocean's volume. The Himalayas, the Alps, and
the Rocky Mountains are all dwarfed by
the Mid-Atlantic Ridge, an enormous range of
mountains that divides the deep Atlantic from
north to south, with only a few scattered islands,
like Iceland and the Azores, peeking above the
waves. The sheer vastness of the deep seafloor
boggles the mind.

Apart from bioluminescence produced by
the animals themselves and perhaps a faint glow
emanating from hydrothermal vents, the deep
sea is dark—much too dark for plants to grow—
so all life in the deep sea consists of either
microbes or animals. Deep-sea forests have
animals instead of trees: sea anemones, corals,
and tubeworms. Deep-sea plains have no grasses
or shrubs, but they do have wandering herds of
animals. Instead of wildebeests and antelopes,
deep-sea herds of sea urchins and sea cucumbers
graze across these plains, their food consisting
not of leaves but of mud. Despite continuous
darkness, high pressure, near-freezing
temperatures, and scarce food, animals occupy
virtually all of the deep seafloor, from Arctic
to Antarctic and from the margins of continents
to the deepest ocean trenches.

Nearly three-fourths of the deep ocean
floor is very flat. Vast abyssal plains lying
between 4000 and 6000 m in the deep ocean
basins are blanketed with the accumulated
skeletons of small plants, protists, and animals
that live and die in the overlying waters. In
relatively shallow seas (less than 3000–5000 m
in most places), the calcium skeletons of tiny
protists (foraminiferans), algae
(coccolithophores), and snails (pteropods) sink
to form soft chalky sediment known
as calcareous ooze. At greater depths, calcium
dissolves away, leaving sediments composed
mostly of silica (glass) skeletons of radiolarian
protists and diatoms. Below the very transparent
oligotrophic waters of the Sargasso Sea and
other midocean unproductive areas, the sediment
has very few skeletons, consisting instead of
fine volcanic ash and desert dust that settles
from the air and sinks below the surface of the sea.

The flatness and apparent monotony
of the abyssal plains belie their biological
diversity. When fine seafloor sediments are
carefully sieved, one finds that most of the
action is below the surface; large numbers of
tiny worms, clams, snails, brittle stars, and
bizarre crustaceans burrow in the sediments or
move slowly through the mud, playing out age-
old dramas of feeding, breeding, and survival.
A careful look at the muddy seafloor from
a submersible window reveals traces of these
dramas in the form of tracks, trails, mounds,
pits, and grooves. Long worms extend their
mucus-covered bodies from deep burrows to
make star-shaped patterns as they sweep up mud
from as far as they can reach. Shrimp, lobsters,
and clams shoot sediment from their holes to
make tiny volcanoes with caldera-like burrows.
On top of the sediment, sea cucumbers by the
millions mop the uppermost layers, licking first
one tentacle then another, while tiny tube-feet

propel them in an endless quest to harvest the bacteria-laden mud. When disturbed, some of these lumbering animals leap off the seabed and dance away with surprising grace. Deep-sea urchins with soft, balloon-like bodies sport whimsical gelatinous bags that hide painfully venomous protective spines. Delicate sponges with roots and skeletons of woven glass reach into the water, filtering bacteria and providing off-bottom resting places for brittle stars, crinoids, and crabs. Small fishes stand quietly on tripod-like fins, capturing whatever the currents deliver. Others hover with heads pointed down, awaiting the emergence of unsuspecting animals from the sediments. Graceful eels, rattail fish, and sharks prowl slowly above the bottom, stalking prey and following the odors of whatever carrion they can find.

Most conditions on the deep seafloor are quite stable, with pressure, temperature, and salinity being virtually unchanged during the course of years, decades, and even millennia. It came as a great surprise, therefore, when scientists discovered not long ago that a small proportion of deep-sea animals reproduce at particular times of the year. How do sea urchins and brittle stars on the deep ocean floor know when it is spring or fall at the surface? Food supply appears to be the answer. Except at hot water vents where chemical energy drives the ecosystem, all animal life on the deep seafloor depends ultimately on the plants and animals that live and die in the upper water column. Dead material from above eventually sinks to the bottom, and because plankton production in the upper water column follows seasonal cycles of light and nutrients, the arrival of dead plankton on the bottom also has a seasonal cycle. Thus, animals gorging themselves on newly arrived plankton corpses (detritus) can invest more energy in egg and sperm production at some times of year than at other times. Once gonads are produced, animals must find mates in complete darkness. Many sea urchins and sea cucumbers wander about by themselves for most of the year, then pair with others of their species for the short breeding season. Nobody knows if they find each other entirely by chance or if they communicate with chemical scents.

The deep seafloor is not entirely flat and muddy. The midocean ridges are made of volcanic rock, as are the active and inactive underwater volcanoes called seamounts. The relatively steep margins of continents are also rocky, and melting icebergs drop stones or boulders onto the ocean floor. Wherever rocks are found, there are specialized animals that live permanently attached to them. Vast reefs of ghostly white corals may extend several meters high off the ocean bottom. Stalked sea lilies (crinoids) bend in the current like tall wind-blown umbrellas, collecting tiny particles from the water that passes them by. Specialized starfishes reach upward with pincer-covered arms that grab the legs of tiny shrimp. Carnivorous sea squirts with wide-open mouths wait for hapless baby fish to land, then grab them like Venus flytraps. Sponges, crinoids, and sea fans extending upward from the bottom are colonized by a diverse assemblage of other animals. In the clear and food-limited waters of the deep sea, high spots with maximum exposure to water currents are prime real estate. The most colorful and diverse assemblages of animals are often found, therefore, on the tops of boulders and the peaks of seamounts. •

Tuscaridium cygneum
Radiolarians
_SIZE 1.2 cm
DEPTH 400–2200 m

Like a constellation in space, these radiolarians, which cannot swim, float in the deep water column. These members of the zooplankton are primitive, unicellular organisms that sometimes form spherical colonies armed with spines. They feed on phytoplankton, but also upon animal prey, such as copepods, jellies, and other gelatinous creatures. After death, their siliceous skeletons rain down to the ocean floor, providing one of the principal components of abyssal sediment. *Tuscaridium cygneum* produce a bioluminescent glow when disturbed.

Bathynomus kensleyi
Giant isopod

_SIZE 40 cm
DEPTH 310–2140 m

It might seem ironic that the largest isopod among the thousands of known species is the one living at extreme depths. This phenomenon is known as abyssal gigantism. Predators are rare in the deep, and animal populations are sparse; if available food resources are scarce, at least they are rather constant. These factors create a relatively stable environment, allowing certain creatures to grow slowly but regularly to sizes much greater than their surface counterparts.

NEXT DOUBLE PAGE SPREAD, LEFT
Umbellula magniflora
Droopy sea pen

_SIZE over 1 m
DEPTH 600–6100 m

The droopy sea pen is a form of coral that anchors in the muddy sediment using a water-inflated bulbous foot. It extends its high stalk to more than a meter over the seafloor so that its feeding polyps can take advantage of currents that are more vigorous than those sweeping right above the bottom.

NEXT DOUBLE PAGE SPREAD, RIGHT
Peniagone gracilis
Sea pig

_SIZE 8.5 cm
DEPTH 200–2500 m

Sea cucumbers contribute more to benthic biomass than any other fauna. The reason for their ecological success is uncertain, but it is probably due to a simple body plan and relatively low metabolic requirements. Sea pigs are not very particular about their diet, satisfying themselves with all the organic particles and bacteria they find in the sediment. After feeding, they sometimes swim away, undulating gracefully toward other pastures.

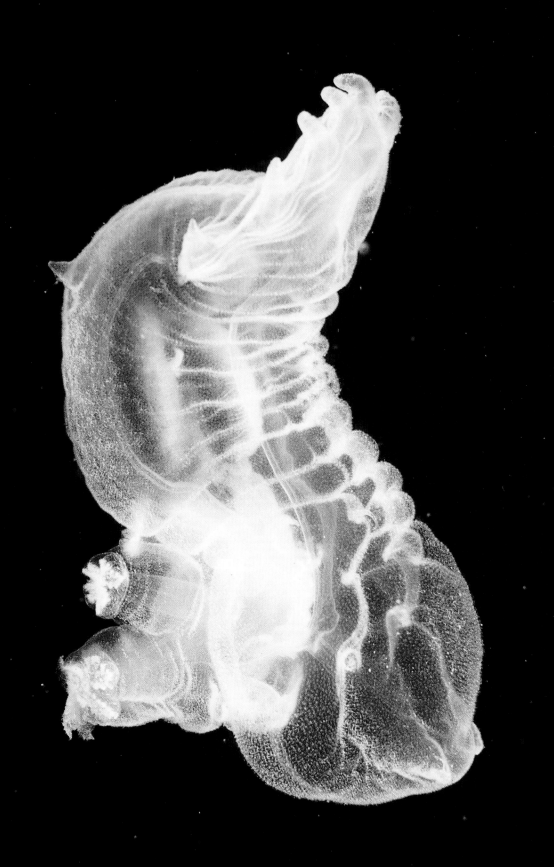

LEFT

Pannychia moseleyi

Glowing sea cucumber

_SIZE 20 cm
DEPTH 212–2598 m

Thanks to its multiple feet, this sea cucumber is highly mobile and can forage over large areas every day. This allows it to have a more selective diet than one would imagine; stomach analyses have shown that this animal favors the richest and most recently deposited organic particles in the sediment. When the glowing sea cucumber is disturbed, its body lights up into a blue-green bioluminescent spiral.

NEXT DOUBLE PAGE SPREAD, LEFT

Histocidaris nuttingi

_SIZE 12 cm diameter without the spines
DEPTH 300–1000 m

The abyssal plain occupies 50% of the ocean floor, an immense expanse of mud across which animals scatter. Certain members of the megafauna, like the *Histocidaris nuttingi* urchins, travel great distances while foraging for their daily nourishment. Still, the real action takes place within the mud rather than on top of it. Depending on the distance from the continent and the depth, between 5 mm and 20 cm of organic detritus deposits on the bottom every thousand years. An entire population of microscopic creatures, the meiofauna, thrives buried within the sediment.

NEXT DOUBLE PAGE SPREAD, RIGHT

Satyrichthys sp.

Armored sea robin

_SIZE 40 cm
DEPTH 50–1000 m

Certain benthic fish spend a large part of their lives alone, in perpetual darkness. The sea robin emits sounds like a birdsong when out of the water. Its flattened form and scarce musculature indicate a sedentary way of life. The front prongs are used for dislodging prey buried in the sediment.

161

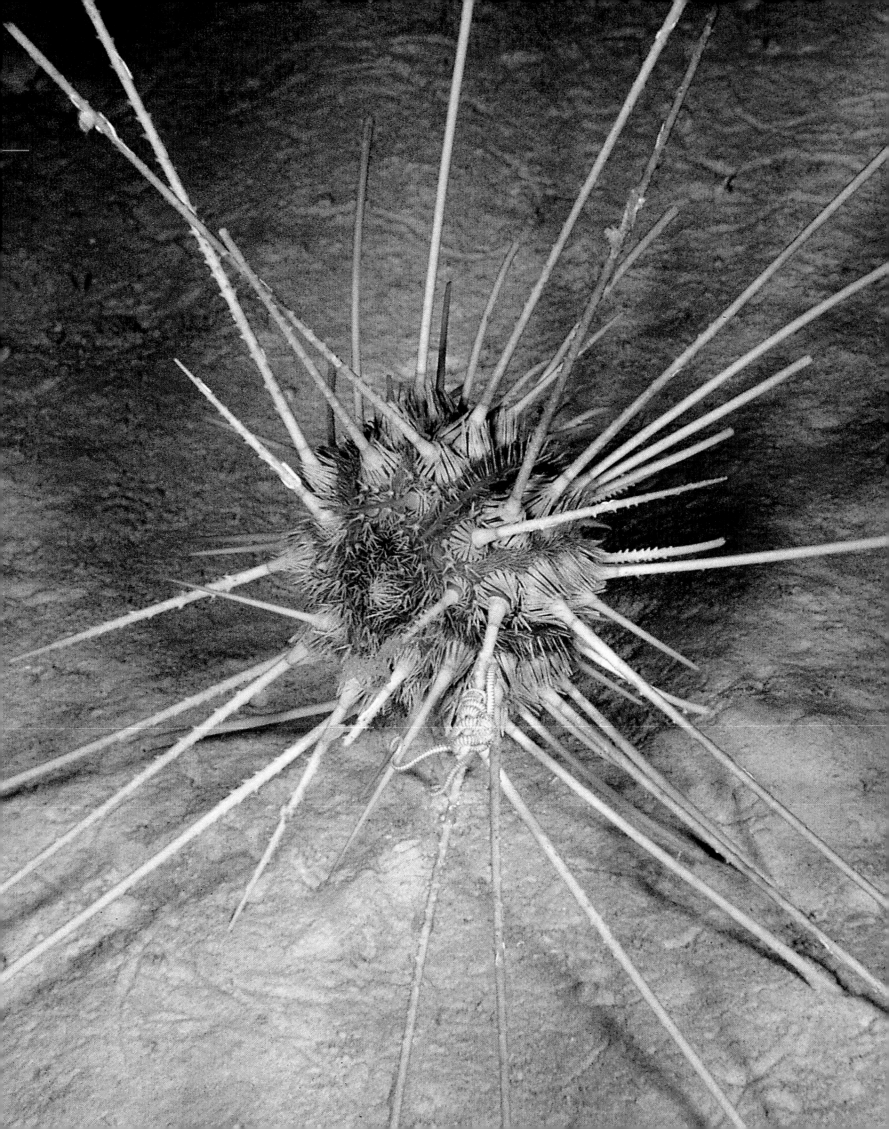

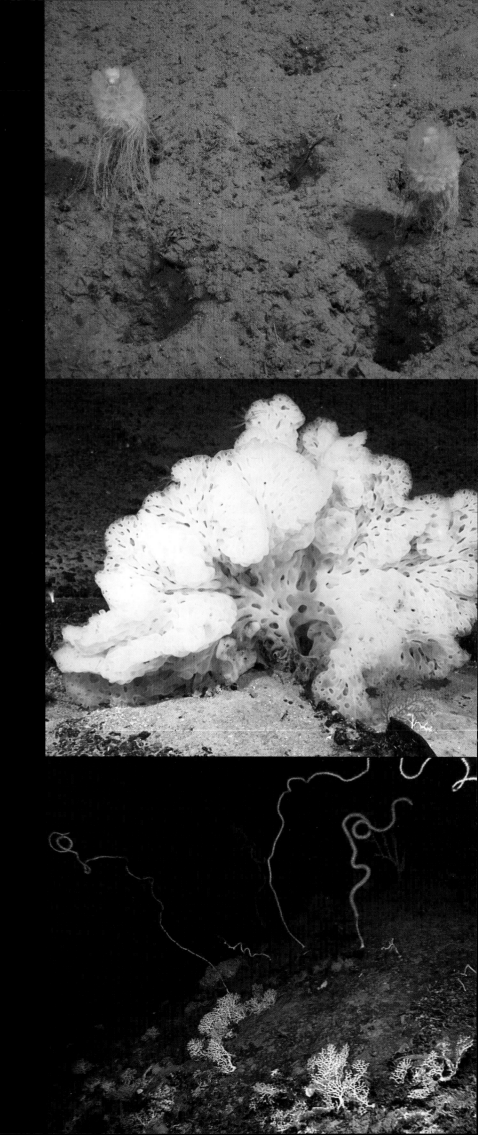

This patchwork of images conveys the diversity of life encountered in the depths. The predatory Venus flytrap anemone (*Actinoscyphia aurelia*, RIGHT PAGE, UPPER RIGHT) awaits a prey. It was so named because of its striking resemblance to the famous carnivorous plant. The remarkable benthic siphonophores (*Dromalia alexandri*, LEFT PAGE, TOP) anchor to the bottom with their tentacles, which gives them the appearance of hot-air balloons ready to take off. Deepwater corals have unexpected shapes, like the 4- to 5-m-high twisted stalks of this undescribed species of bamboo coral (LEFT PAGE, BOTTOM) or the tall spiraling coral *Iridogorgia* (RIGHT PAGE, LOWER LEFT). Sponges number among the very first creatures to have appeared on the planet. In the deep, the delicate glass sponges, which sometimes display astonishing colors (hexactinellid, LEFT PAGE, MIDDLE), can reach a meter in height after growing for two hundred years. Among the mobile fauna, the tripod fish (*Bathypterois dubius*, RIGHT PAGE, MIDDLE RIGHT) is symbolic of deep-sea life; the spokes of its pectoral fins have evolved into crutches that lift the fish into the current. The slender frostfish (*Benthodesmus tenuis*, CENTER) is seen maintaining a curious position for a fish; though this two-meter giant is probably just resting...

OPPOSITE

Ptychogastria polaris

_SIZE body 7 cm without the tentacles
DEPTH 0–3000 m

This beautiful jelly is usually
a deep-dwelling species, though
it can be seen near the surface
in the Arctic and Antarctic. Animals
found at both poles are rare;
more commonly, species are found
at one pole and in the deep.

The Arctic Ocean is a realm of magical contrasts.
Where else on Earth could a giant bear and
an ethereal abyssal octopus coexist peacefully
at precisely the same map coordinates? Or
a jellyfish normally found thousands of meters
deep be encountered by scuba divers? We tend
to picture the high Arctic as an icy landscape, but
this is only half the story. When Robert Peary—
the first explorer to reach the North Pole—
achieved his goal in 1909, he stood on a layer
of ice 3 m thick; beneath that frozen crust lay
more than 4300 m of liquid ocean.

It is hard to imagine a tougher place
for an animal to eke out an existence than the
frigid deeps beneath the polar ice. But a surprising
variety of deep-sea life has adapted to these
conditions, which are not as challenging at
they might at first seem. In fact the environment
deep beneath the North Pole is not much different
from that found in the deep sea beneath
the equator—it is cold, it is dark, the pressure
is immense, and food is scarce. Many of
the same animal groups, and even species,
exploit both habitats.

The existence of life in the Arctic abyss
has been known for a surprisingly long time.
In 1818, twenty-three years before Edward
Forbes famously proposed that the deep sea
was devoid of life, the English explorer John
Ross accidentally captured a basket star
(*Astrophyton*) from Arctic waters 1600 m deep
as he was sounding the depths in his quest
to find the near-mythical Northwest Passage.

Typically, however, the Arctic Ocean
doesn't yield its secrets so easily. It is by far
the least known of the world's oceans and is

Dr. Michael Klages
Alfred Wegener Institute,
Germany

THE POLAR DEPTHS

the most isolated from its neighboring seas.
Surrounded by land, the Arctic's only deepwater
connection to the rest of the world's deep oceans
is the Fram Strait, between Iceland and
Greenland. The ocean itself is divided into
several deep, flat-floored basins covered with
fine mud separated by ocean ridges—one of
which, the 1800 km–long Gakkel Ridge, is part
of the global midocean ridge system. It has
the slowest spreading rate of any ocean ridge:
0.3–1 cm per year compared to about 6 cm
per year elsewhere. A very special feature of
the Gakkel Ridge is the great depth at which it
lies: about 5000 m—twice as deep as most other
ridges. Contrary to former assumptions that
there would be little or no hydrothermal activity
along such ultraslow-spreading ridge systems,
we now know that it actually might be stronger
than at any other ocean ridge yet studied.
Because of its quasi isolation from other ridges,
the Gakkel Ridge is almost certainly home to
undiscovered species of hydrothermal vent life.

As is true elsewhere in the deep ocean,
the greatest challenge for life in the deep
Arctic basins is the sparse food supply. During
the intense, near-continuous sunlight of summer,
microscopic algae thrive within and under the
Central Arctic ice sheet, and these tiny organisms
are the ultimate source of sustenance for life
on the abyssal plain, kilometers below.
The fraction of this meager food supply that
reaches the Central Arctic seafloor is vanishingly
small, and in places this never-ending famine
restricts the biomass of bottom life to less than
a gram per square meter. Elsewhere in the Arctic,
the ice sheet is not a permanent feature,
and productivity is much higher. If the latest
climate change predictions are true, by 2100
the entire ice cap will disappear in summer,
with enormous consequences for the Arctic abyss.

Only recently have the true diversity
and density of life under the ice been revealed.
Above the bottom, hanging in watery space,
can be found a marvelous variety of plankton.
In the Central Arctic biologists have identified
more than a hundred species of crustaceans,
more than sixty jellies, and ten species of
arrowworm. Curiously, several of the jellies can

be found from near the seafloor all the way up to the surface. These species also occur at great depths in warmer seas, and it seems likely that the relatively cold, dark conditions found near the surface in the Arctic replicate their more usual deep-sea habitat.

On the Arctic seafloor live many of the deepwater groups that characterize abyssal plains everywhere—echinoderms, worms, bivalve mollusks, and sponges. Larger animals are rare in the Arctic abyss, although at least half a dozen species of octopuses and squids make the deep Arctic their home, as do a number of fishes. The largest of these is the Greenland shark, *Somniosus*, a sluggish 6 m giant.

At first sight the oceans of the Arctic and Antarctic seem similar, but one could say they are poles apart. As biologist Andrew Clarke points out, the differences outweigh the similarities. Antarctica is an ancient continent centered on the South Pole and surrounded by ocean—a kind of photographic negative of the Arctic situation, where the ocean is surrounded by land. The Arctic abyss is geographically isolated from the other oceans, whereas the Southern Ocean has deep connections with the Atlantic, Indian, and Pacific.

If the Arctic is notable for its abyssal basins, the Antarctic is best known for its unusually deep continental shelves. These are the shallow, offshore extensions of every continental landmass; throughout most of the world they are normally 100–200 m deep.

But the shelves of Antarctica are up to 800 m below sea level. This is because the shelves are part of the Antarctic landmass, which is pushed down in its entirety by the weight of the kilometers-thick layer of ice that formed above it. As the ice sheet grew in thickness, the shelves sank into the ocean like an overloaded dinghy.

The waters fringing Antarctica are seasonally productive, although overall the Southern Ocean is not as bountiful as is often claimed. Food availability varies intensely throughout the year, and many animals have evolved ways of dealing with this feast and-famine way of life. Antarctic krill, one of the Southern Ocean's great biological successes, actually shrink in size during winter—an extraordinary feat for a hard-shelled crustacean.

During repeated glacial periods, the ice volume on the Antarctic continent increased, leading the shelf ice to cover the seafloor down to 600 or more meters, thus destroying all benthic life. The only option to survive under such conditions was to go deeper, forcing the animals to adapt to a changing habitat. This phenomenon, known as the "diversity pump," caused the evolution of new deepwater species, which then spread northward into the other oceans. So the polar zones may be more than simply an interesting backwater in the global abyss; they could be the cradle of deep-sea life—the deep ocean's equivalent of Africa's Rift Valley, where humans are thought to have taken their first steps. •

OPPOSITE

Desmonema glaciale
_SIZE 5 m
DEPTH unknown

Desmonema is an immense jelly whose bell can reach a meter in diameter and whose long ribbon-like tentacles may be up to 5 m. Jellies and ctenophores dominate the pelagic fauna of the Antarctic in number and variety.

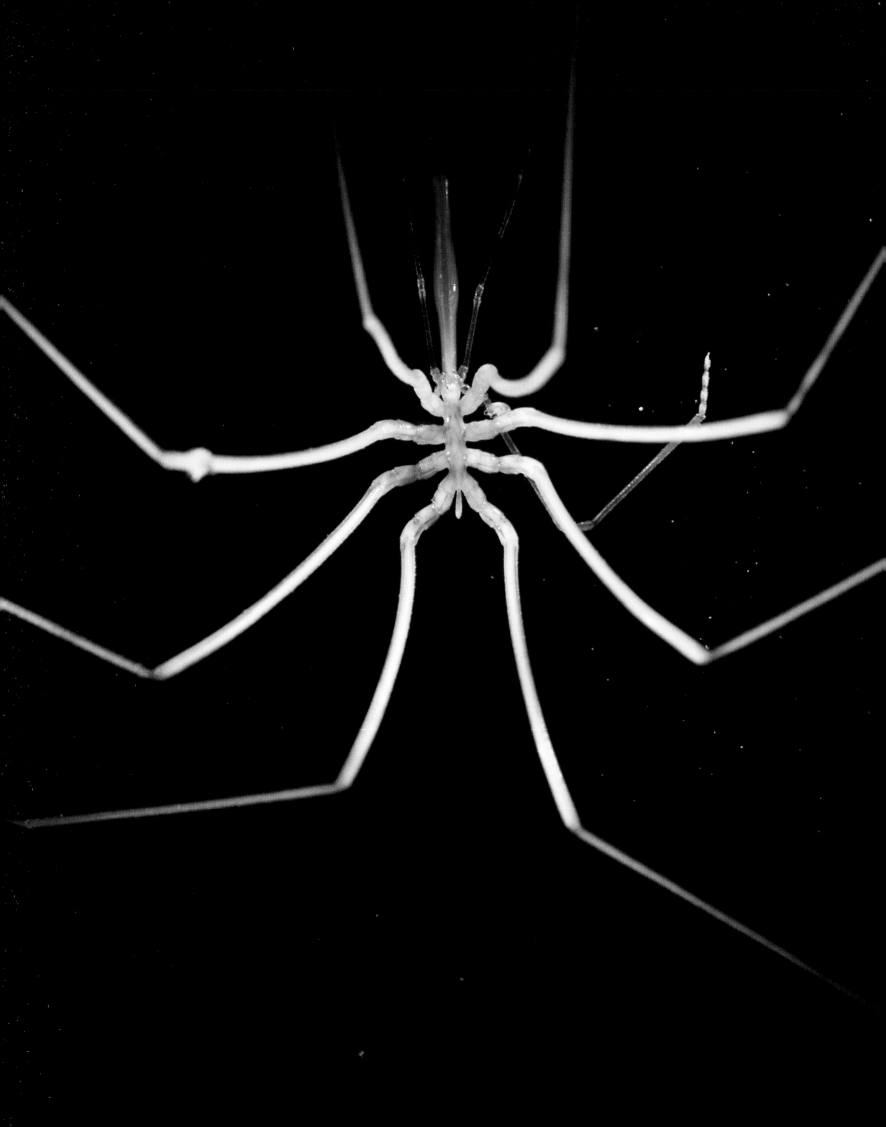

OPPOSITE
Unidentified species
_SIZE 20 cm leg span
DEPTH unknown

TOP AND MIDDLE
Colossendeis sp.
Giant sea spider
_SIZE 30 cm diameter
DEPTH 10–7400 m

BOTTOM
Ammothea verenae
Vent pycnogonid
_SIZE 6 cm
DEPTH 1500–2400 m

Pycnogonids can be found at all ocean depths, but only the abyssal species and their close cousins living at the poles reach giant sizes, with a leg span of over 50 cm. Their immensely long appendages make them very mobile, but they are surprisingly slow. They search the bottoms for soft, stationary meals that cannot escape, like anemones. They suck the tissues out of their prey using a proboscis that they drive into the creature like a straw into a milkshake. Pycnogonids are generally solitary, but at the foot of hydrothermal chimneys, one small and blind species (*Ammothea verenae*) can be seen in dense aggregations. These are probably reproductive colonies where the males collect the eggs of several females and incubate them on their abdomens. The white, cottony flakes are bacterial filaments that the sea spiders feed on.

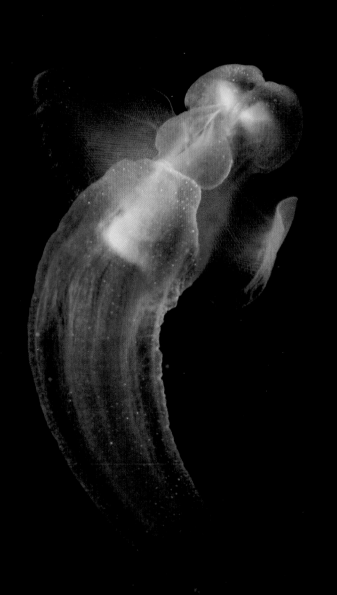

PRECEDING DOUBLE PAGE SPREAD, LEFT

Clione limacina

Naked sea butterfly

_SIZE up to 8.5 cm
DEPTH 0–600 m

Although it can fast for an entire year, the naked sea butterfly proves itself an agile and capable hunter when resources abound. It feeds exclusively on other pelagic gastropods, which it extracts from their shells with large hooks and swallows whole, in keeping with its somewhat barbaric table manners.

PRECEDING DOUBLE PAGE SPREAD, RIGHT

Actinostola callosa

_SIZE 30 cm in height
DEPTH 10–1480 m

Certain Norwegian fjords are characterized by steep slopes that plunge rapidly beyond the photic zone. Numerous representatives of deep oceanic fauna are found there, like the *Actinostola callosa* anemone and one of its favorite prey, the *Periphylla periphylla* helmet jelly. The latter is a delight to the anemones that, with such an abundant prey, have also become very widespread.

LEFT

Promachocrinus kerguelensis

Sea lily

_SIZE 50 cm spread
DEPTH 10–2100 m

There are no currents immediately above the substrate, though they increase rapidly as one rises from the bottom; this is why any roosting spot, whether mineral or animal, like the sponge being used here, is highly prized by creatures of the deep. Even a few centimeters of elevation improves the probability of capturing food suspended in the water. This sea lily, the largest and most abundant in the austral oceans, intercepts them with its arms deployed like a crown. Small cilia then transport the nutrients to the mouth, situated at the center of the animal.

"I had a profound conviction that the land of promise

for the naturalist, the only remaining region where there were

endless novelties of extraordinary interest ...

was the bottom of the deep sea."

Sir Charles Wyville Thomson, 1872

PRECEDING DOUBLE PAGE SPREAD, LEFT
Periphylla periphylla
Helmet jelly
_SIZE up to 1 m
DEPTH 0–7000 m

The helmet jelly is usually found at
very great depths, but living
conditions at the poles are close
enough to those of the abyss that it
is possible to observe a certain
number of these creatures at the
surface there. *Periphylla periphylla*
is very likely the most widespread
deep-sea jelly in the world.

PRECEDING DOUBLE PAGE SPREAD, RIGHT
Unidentified species
_SIZE 30 cm
DEPTH 800 m

Ever since the first polar expeditions
of the nineteenth century, some very
curious cephalopods, usually
confined to the great depths, have
been discovered at the surface of
the extreme latitudes. These
denizens of the deep somehow get
lost in the cold currents circulating
between the poles and the bottoms.
Except for the luminosity, the polar
environment is very similar to
the abyssal environment, allowing
certain deep-sea creatures to dwell
there without any problem.

OPPOSITE
Latrunculia apicalis
Green globe sponge
_SIZE 12 cm height
DEPTH 10–1200 m

Sponges have retained through
the ages a simple body plan that
consists of an assembly of filtering
cells that grow very slowly and
without a precise pattern, which
gives rise to a variety of forms.

Megalodicopia hians
Predatory tunicate
_SIZE 20 cm
DEPTH 180–1000 m

Near the surface tunicates, just like sponges, filter water and retain the nutrients they need in the process of this respiration. The predatory tunicate has developed a unique strategy: it is both a filterer and a predator. Its large mouth closes rapidly on small shrimps or other crustaceans that stray there.

Imagine yourself near the head of the Grand Canyon in Arizona. You are looking up at the canyon walls, observing the colorful organized sedimentary beds that provide a tapestry of motives and texture. Now, imagine that you are standing in the same place, but this time under 200 m of water with the light considerably reduced so that all you can see is fluid blues and greens with bladed shafts of light piercing the darkness from above. Looking down canyon you see nothing but blackness. Around you, the walls have a dusting of sediment much like you would find in an old abandoned house; scattered about them, a mosaic of filigree and filamentous organisms gently sway with the current. If you look away from the walls you may see sharks and rays slowly swimming by, silently sizing you up. You think this vision was taken out of an underwater fiction? No, this scenery is real and it can be seen down in the Monterey Canyon off central California.

The Monterey Canyon is one of the larger submarine canyons found along the western coast of the North American continent. It is equal in size to the Grand Canyon, but it is so completely covered by water that there is no hint of its existence from above the sea. The same is true of all other underwater canyons around the world: they simply cannot be spotted from above. It took the advent of underwater exploration vehicles in the 1960s and '70s to start observing them firsthand; it turned out that submarine canyons are dominant habitats for fishes and other organisms along the coastlines of the world's oceans.

Dr. Gary Greene
Monterey Bay Aquarium Research
Institute (MBARI), USA

These spectacular features have various geological origins: they can either be land canyons that were submerged during times of sea level rise, ice-carved canyons that were eroded from glaciers during times of sea level low stands, or canyons formed by submarine landslides. Some of the most impressive submarine canyons are located off the mouths of large rivers such as the Nile River in Egypt. A lot of what we know of them is drawn from the knowledge that the Monterey Bay Aquarium Research Institute has been accumulating on what is certainly the best-studied canyon in the world today: the Monterey Canyon.

The Monterey Canyon is a 170 km–long canyon whose floor reaches a depth of 3800 m. Its geological formation is still under debate: geologists agree the canyon was formed underwater but its origin remains a mystery in that it is not connected to a river system that has the capability to erode its deep granitic basement axis. The present and leading hypothesis argues that the Monterey Canyon formed some time between ten and fifteen million years ago and has moved into its present location from the Santa Barbara area, some 350 miles to the south, by plate motion.

Not many canyons are as sinuous as the Monterey Canyon, which is made up of many meanders and side canyons. This unique aspect comes from the fact that it sits within the plate boundary that separates the Pacific Plate from the North American Plate. These plates move against each other at a rate of about 8 cm per year, the rate at which fingernails grow, and the tremendous force resulting from this movement has deformed and fractured the rock here. Earthquakes are common along the faults of the plate boundary, and this shaking often produces landslides, which dislodge chunks of rocks off the steep walls of the canyon. These fresh rock walls provide hard substrate for encrusting organisms such as sponges and refuge for fishes and other mobile fauna.

Some of Earth's most diverse and unusual biology is found in submarine canyons. Their biological specificity lies in their morphology and

THE MONTEREY CANYON

geographic situations along continental margins. As opposed to continental slopes that progressively lead towards the deep parts of the oceans, canyons cut deep across the continental shelf perpendicular to the coastline and provide rapid access to the deep ocean. They are natural conduits for the landward transport of deep, nutrient-rich oceanic waters.

These waters started in the polar region as surface waters that sank to the bottom and slowly moved along the seafloor until they reached the Monterey Canyon and came back up to the surface in a process known as "upwelling." Because the deep sea receives the carcasses of small organisms that lock up nutrients during photosynthesis and release them when they die, the deep waters are extremely rich in nutrients. When the water returns to the surface after a long enriching deep journey, it triggers an explosion of life with plankton blooming thanks to the nutrients, and in turn, a whole chain of predators feeding off the planktonic base.

These highly productive waters affect the whole canyon ecosystem. Animals come in all shapes and forms: as encrusting organisms on the canyon walls, as filter feeders attached to the rock outcrops, or as pelagic animals cruising around the canyon's many meanders. Mammals such as orcas and whales often concentrate in submarine canyons to feed on the plentiful plankton. Because of the enhanced water circulation due to the topography, canyon walls are strategic places for organisms that rely on currents to bring them food. Impressive deepwater corals and gorgonians, rarely encountered on the deep seafloor, thrive here. Other animals generally restricted to the deep ocean can be found in the shallower parts of the canyon. A suite of organisms comprising deep-sea jellyfishes, mollusks, and fishes or cephalopods benefit from the presence of all the food raining down from the surface waters. With such abundant resources, canyons also act as reproduction and spawning grounds for certain

animals like deepwater sharks, which come here to attach their egg cases and increase the survival chances of their offspring.

The sediment layer on the canyon edges is dramatically enriched by the high biomass production. Walls are coated with mud, drift kelp, sea grasses, and pelagic material made of gelatinous organisms that thrive in the sunlit waters above. This organic-rich sediment is frequently carried down to the deep sea either by landslides stimulated by a seismic event or by extremely fast downhill currents triggered by the weight of suspended elements. These underwater turbulent cascades, called "turbidity currents," are very common in submarine canyons; they can travel hundreds of kilometers and contribute to bringing food to the deep sea.

Since organic-rich sediment is deposited deep in the canyon, decomposition rapidly occurs, using up much of the available oxygen and thus producing methane and sulfide-rich fluids that provide the chemistry needed to support sulfide-dependent animal oases of bacteria, clams, and worms. Massive amounts of these chemosynthetic clams have been found in the Monterey Canyon at a depth of around 800 m, called clam acres, and produce a spectacular carpet of yellow-brown color that contrasts with the green gray mud that commonly carpets the canyon axis.

The Monterey Canyon is an ideal place to make deep-sea biological observations: it is close to shore yet extremely deep and the exceptional productivity of its surface waters leads to an equally exceptional concentration of deep-sea organisms, especially when compared to deep-ocean basins, where animals are spread out kilometers apart. The sinuosity of the canyon prevented it from being thoroughly explored using traditional methods such as trawl nets, but the recent development of high-tech tools such as ROVs (remotely operated vehicles) changes the face of exploration and helps reveal the numerous remaining mysteries of the Monterey Canyon. •

OPPOSITE
Ptychogastria polaris
_SIZE body 7 cm, not counting tentacles
DEPTH 0–3000 m

Jellies are usually thought of as creatures that spend their entire lives in the midwater, far from sharp reefs or rocks. *Ptychogastria polaris* is an exception. Though it is quite capable of swimming, it is most often found attached to rocky walls with its small adhesive tentacles. It deploys other, longer tentacles for capturing prey in the currents.

NEXT DOUBLE PAGE SPREAD
Computer-generated image

For nearly twenty years, the remotely operated vehicle *Ventana* has been diving daily in the sinuous Monterey Canyon in California, at nearly 2000 m depth. Before the Monterey Bay Aquarium Research Institute began studying the mysteries of the deep, very little was known about submarine canyons.

PAGE 186
Solmissus sp.
Dinner plate jelly
_SIZE 20 cm
DEPTH 700–1000 m

This jelly is widespread at all depths of the Monterey Canyon, feeding principally on large, gelatinous prey.

PAGE 187
Chondrocladia lampadiglobus
Ping-pong tree sponge
_SIZE 50 cm height
DEPTH 2600–3000 m

This curious organism, which looks like a 1960s chandelier, is a sponge! In contrast to its relatives that filter water and retain nourishing particles, this sponge is carnivorous. Creatures that alight on its surface are captured by tiny, filamentary hooks, called spicules, which act like a Velcro strip. This triggers a reaction in the cells of the sponge, which migrate toward the prey in order to digest it. Once the small crustacean or worm has been consumed, after several days, the cells resume their initial position.

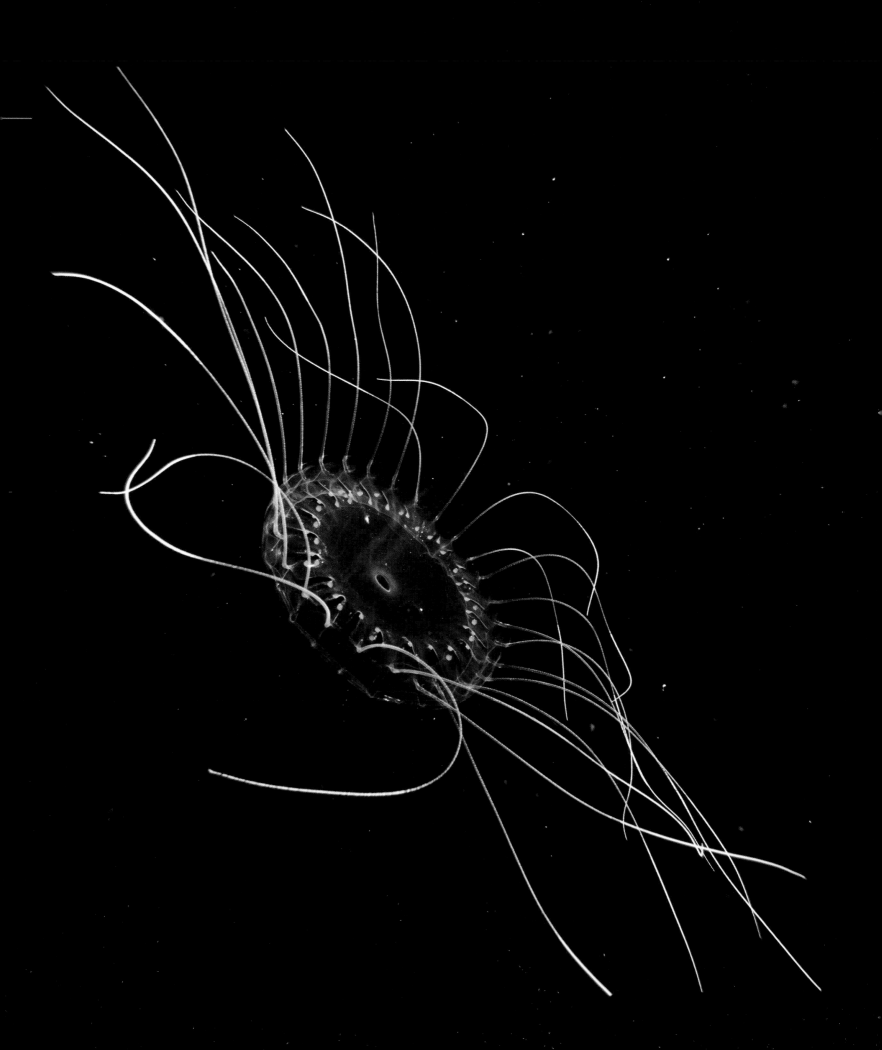

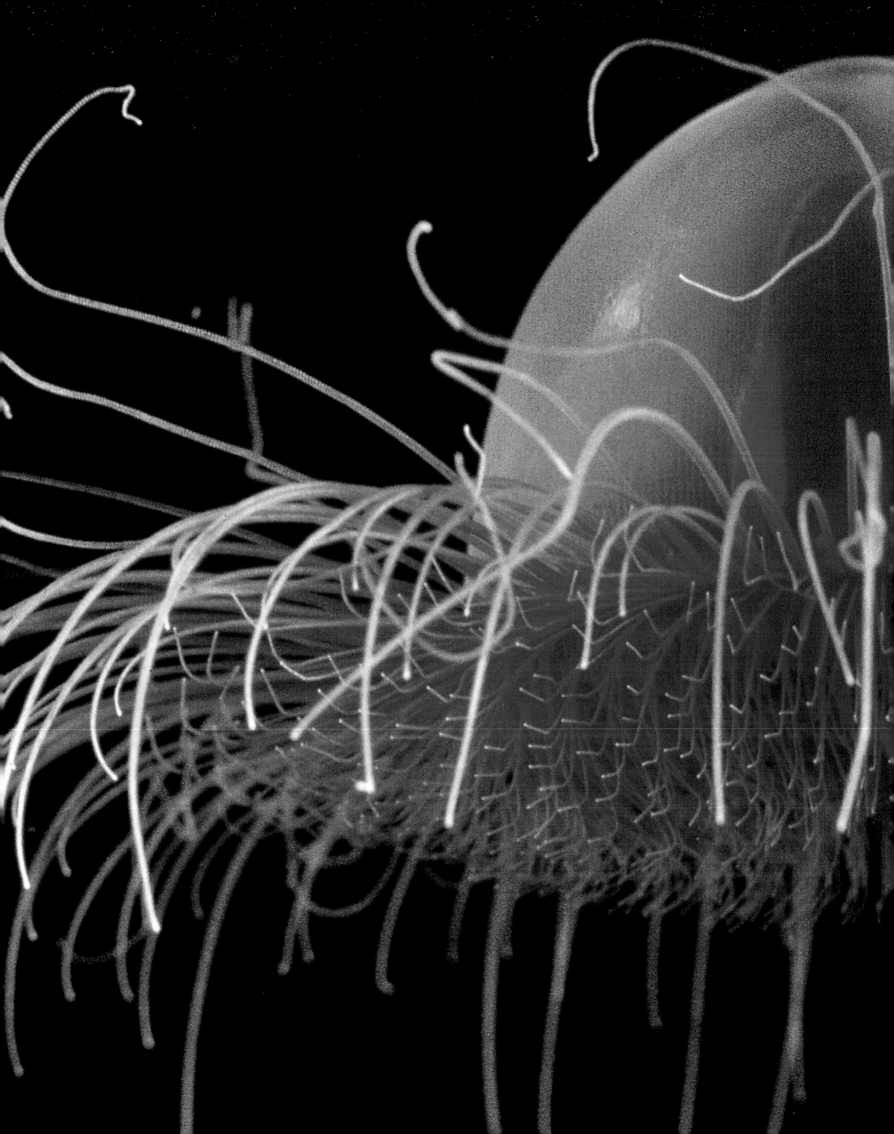

One most often thinks of sharks as rapid predators that navigate exclusively in the surface waters, but in reality, about 60% of the world's four hundred known shark species live at the slow-motion rhythm typical of the oceans' greatest depths. Only a very small number of these deep-sea species migrate toward the surface at night. Accordingly, the sharks of the dark remain highly enigmatic.

We do know that their dietary regime can be described as opportunist, in that they seize upon any chance whatsoever for nourishment, whether in the form of living prey or carrion. The sleeper shark (*Somniosus* sp.), for example, has been filmed ripping off huge chunks from the carcass of a whale.

Stomach analyses of this great, lethargic 4 m–long shark might even make one think they feed exclusively on dead whales. The sleeper shark is one of the rare sharks that have been observed in their natural environment; in fact, the majority of the others are known only from dead specimens brought in by trawlers. In order to understand the behavior of these creatures of the deep when they can't be observed directly, one needs to put on the detective's cap. In the case of the cookie cutter shark (*Isistius* sp.), the exact match between the bite marks on the skin of its targets— large fish and whales—and the jaw of this small, half-meter-long shark was the key to the mystery of the "cookies." This little shark's table manners are typical of the family of Dalatiidae sharks: they sink

SHARKS OF THE DARK

their teeth into their prey and then spin their body around until they've twisted off a chunk of flesh. This technique leaves circular marks with jagged edges that recall the appearance of some American cookies. Large sharks, tunas, and even whales go through life scarred by these encounters.

The sleeper shark's size is remarkable because most deep-sea sharks are smaller than their cousins living at the surface. There are a few notable exceptions; the megamouth shark (*Megachasma pelagios*) can reach a length of 5 m. Dogfish (*Centroscyllium* sp., *Squalus* sp., and *Centroscymnus* sp.), which can measure between 90 and 160 cm, are the most highly exploited deep-sea sharks today. They are sought for the high concentrations in

their livers of squalene, an oil that is used primarily in the cosmetic industry. Additionally, there is an Asian market for shark fins and tails. We already find their meat, cut up into filets, as a regular part of our diet; this is because of the drastic impoverishment of traditional surface resources. Sharks of the deep have a slower metabolism and begin reproduction late; their gestation is quite long and produces few young. The combination of these factors makes their exploitation unviable. Moreover, it is estimated that their population has dropped by 80% in the Northern Atlantic since 1992. Some scientists believe that sharks are actually the most vulnerable of all animals of the deep, and the most endangered, unless fishing can be strictly regulated in the very near future. ●

PRECEDING DOUBLE PAGE SPREAD
Voragonema pedunculata
_SIZE 4 cm diameter
DEPTH 500–3500 m

The myriad tentacles (between 1000 and 2000) of this benthic jelly are used to capture small crustaceans.

OPPOSITE TOP AND ABOVE
Chlamydoselachus anguineus
Frilled shark
_SIZE 2 m
DEPTH 0–1600 m

Recent studies have shown that the females of this eel-like shark carry their young for three and a half years, a gestation period nearly twice as long as that of elephants. When it surfaces from the depths to give birth, it usually chooses the nutrient-rich shallows of submarine canyons.

LEFT
Callorhinchus milii
Elephant fish chimaera
_SIZE 1.25 m
DEPTH 0–227 m

Because of its rabbit teeth, elephant trunk, and silvery color, one can understand the name of the elephant fish chimaera, after the monstrous hybrid in Greek mythology. This close relative of the sharks uses its fleshy protuberance to work the sediment in search of bivalves that it grinds with its powerful, flattened teeth. In spring, females lay 25-cm-long eggs, which take between six months and a year to hatch.

NEXT DOUBLE PAGE SPREAD
Anthomastus ritteri
Mushroom soft coral
_SIZE up to 15 cm
DEPTH 200–1500 m

With all its arms gathered back into its body, this soft coral forms a perfectly round ball at the end of a massive stem, lending it the appearance of a mushroom. Its color varies from white to red, passing through all sorts of orange and rose tints. Some shark species lay their eggs amid the polyps, which contain a toxin that protects the vulnerable eggs from predators.

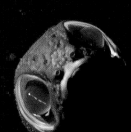

Bathylagus pacificus
Slender blacksmelt

_SIZE 25 cm
DEPTH 230–7700 m

Researchers have nicknamed this small, primitive fish with a sulky expression the owl fish because of its eyes. The fish is adapted to the extreme depths; it has no swim bladder (the air-filled sac that most fish have for maintaining neutral buoyancy), its skeleton is very light, and its leathery-looking skin has no scales. The deeper one goes, the less calcium is in the water, and so more energy is required for constructing a skeleton or scales. The slender blacksmelt has a different tool for effortless floating: a subcutaneous layer of gelatinous matter lighter than water. Because of its small size, it has no commercial value, in contrast to other species populating the areas around submarine mountains.

In January 2005, when a U.S. submarine crashed into a 2000 m–high seamount rising to within 50 m of the surface near Guam, in the Western Pacific, it seemed incredible to many that the feature was uncharted. But the deep sea is more poorly mapped than Mars or the far side of the moon. Scientists remain uncertain whether the number of seamounts taller than 1000 m is closer to 25000 or 50000 in the world's oceans, and the number of smaller seamounts is variously estimated to lie somewhere between 100000 and 1.5 million.

Seamounts are submerged mountains, usually of volcanic origin. The vast majority are found in the Pacific, associated with that ocean's many island chains, submarine ridges, and other volcanic features. But if seamounts today remain largely uncharted, their influence on the diversity of the life in the oceans is even more poorly understood.

Seamounts greatly amplify the ambient deep-sea currents, much as mountains on land create updrafts and influence the airflow around them. These currents, in turn, create a unique deep-sea environment. They serve first to winnow away the sediments on the seamount summit and slopes, leaving the bare rock substrate exposed. In addition, they enhance the flow of prey organisms over the seamount. As a result, in marked contrast to the sediment-covered plains of the deep sea, sparsely populated by worms and other small burrowing organisms,

seamount substrates are often colonized by a variety of soft and hard corals, sponges, crinoids, anemones, and a great variety of other suspension feeders that remove small prey swept past by the currents. Seamounts are also often home to large schools of fishes, such as orange roughies, pelagic armorheads, oreos, and alfonsinos. Quite unlike the weakly swimming fishes typical of the deep sea, seamount fishes have robust bodies and are adapted to maneuvering within the strong seamount currents. Typically, they feed on the larger prey—small fishes, squids, and prawns—that drift past the seamount or are trapped over the relatively shallow topography when they attempt to swim back down to greater depths during their diurnal migrations.

Currents tend to be guided by the underlying submarine topography, so the currents sweeping over seamounts often form an eddy or gyre, known as a Taylor column, that can extend hundreds of meters above the seamount. This circulation cell, slowly revolving over the seamount, can retain the eggs and larvae of seamount populations. The Taylor column can also raise or "upwell" nutrient-rich deep water toward the surface, promoting increased productivity of the microscopic algae at the base of ocean food webs. As a result, seamounts far beneath the surface are often hotspots of near-surface marine life, attracting seabirds, sharks, and tunas, as well as deepwater life forms.

Deep-sea corals, unlike those on tropical reefs, do not live in association with photosynthesizing algae. Dependent on enhanced currents to bring them adequate prey, they are largely restricted to seamounts, canyons, and certain continental margins, where these currents prevail. Though less well known than their warm-water cousins, there are in fact more species of hard corals now known from deepwater habitats than from tropical reefs.

Seamount environments are rugged and difficult to sample, so they remained poorly studied until quite recently. Unfortunately,

Dr. J. Anthony Koslow
The Commonwealth Scientific & Industrial Research Organisation (CSIRO), Australia

seamounts:
galápagos of the deep

fishermen were the first to discover the profusion of life on seamounts rather than scientists. In the late 1960s, Soviet trawlers first discovered large aggregations of pelagic armorhead living over the seamounts of the Emperor-Hawaiian seamount chains northwest of Hawaii. Aggregated on these relatively small topographic features, seamount fishes are highly vulnerable to modern trawling. Within ten years the Soviets, later joined by the Japanese, trawled almost a million tons of armorhead from the region, rendering the fishery commercially extinct. The Soviet trawlers then went on to discover orange roughies on seamounts around New Zealand. At about the same time, Japanese fishermen discovered precious pink and red corals on the Emperor seamounts, which led to a rush by more than a hundred vessels to strip and harvest the corals, used in jewelry making since antiquity, from Pacific seamounts. Today seamounts have been fished in all the world's oceans. Most such fisheries follow a boom-and-bust cycle and are typically depleted within five to ten years. Some seamount fishes, such as orange roughies and oreos, can live to more than a hundred years and require twenty to thirty years to attain sexual maturity. These species are adapted to the relatively quiescent conditions of the deep sea, where they have few predators and mortality rates are extremely low. Modern industrial fisheries on such species are clearly not sustainable.

Corals and other benthic fauna are a bycatch of trawling on seamounts; it removes them along with the fish. Scientists are now finding that deepwater coral reefs are home to hundreds of species of fishes and invertebrates, much like shallow tropical reefs. Expeditions to explore the seamount faunas of the Southeastern Pacific and in the Tasman and Coral Seas have found that 25 to 50% of the species are new to science and apparently endemic, occurring only on seamounts in these regions. Some seamount corals and sponges have been found to live to one hundred or even several hundred years, and extensive deepwater reefs of the coral *Lophelia* off Norway date back to the end of the last Ice Age, about ten thousand years ago. The high rates of endemism on seamounts, very unusual in the deep sea, probably arise from the tendency of the currents to follow the seamount and ridge topography, thereby restricting the dispersal of eggs and larvae. In this way seamount clusters become reproductively isolated from one another, allowing for the evolution of local species groups, much the way that small islands, like the Galápagos, may evolve species of birds and reptiles distinct from those found on the rather distant continent. Unfortunately, their restricted distributions leave them vulnerable to extinction from intensive trawl fisheries. Australia, New Zealand, the U.S.A., Norway, and other countries have now protected seamount and deepwater coral areas within their jurisdictions from trawling. The UN General Assembly passed resolutions in 2002 and 2003 that recognized the need to protect the biodiversity of seamounts on the high seas—areas beyond national jurisdiction—but concrete steps to protect them have not yet been taken. •

OPPOSITE
Lophelia pertusa
Tuft coral
DEPTH 10–2500 m
When abyssal tourism gets up and running, the crowds will doubtless flock to admire the remarkable beauty of the deep coral reefs—or rather, what remains of them, because fishing trawlers have already destroyed 50% of these ecosystems in certain reefs, like those off the coasts of Norway. With their maximum growth of 2.5 cm per year, deepwater corals are truly vulnerable.

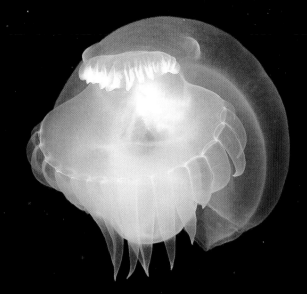

Submarine mountains present striking relief in the midst of the abyssal plain. All sorts of creatures, whether mobile or sessile, come there to feed, or to reproduce, taking advantage of the increased currents. These geological formations, mostly extinct volcanoes, along with a few that are still active, are referred to as diversity hotspots or undersea islands. Whether in the Gulf of Alaska, in the South Pacific, or off the coast of California, researchers discover new wonders with each dive: gigantic coral forests, like those formed by the 2 m–high bubblegum coral (*Paragorgia* sp., RIGHT PAGE, UPPER LEFT); the frail but immense bamboo corals several hundred years old (RIGHT PAGE, LOWER RIGHT); sponges with chemical substances that hold promise for medicine; or pelagic creatures attracted by the wealth of resources. Among these latter are the small *Benthocodon* sp. jelly (LEFT PAGE, BOTTOM) and the famous "mystery mollusk" (LEFT PAGE, TOP) discovered by the Monterey Bay Aquarium Research Institute at 2500 m depth and which, even today, defies classification among the mollusk groups known by researchers. ROVs or submersibles encounter some exceptional creatures, such as the unknown animal, doubtless a bigfin squid (*Magnapinna* sp., CENTER), with tentacles 5 m long forming a very unusual angle among cephalopods. This giant cruises the oceans at 5000 m depth.

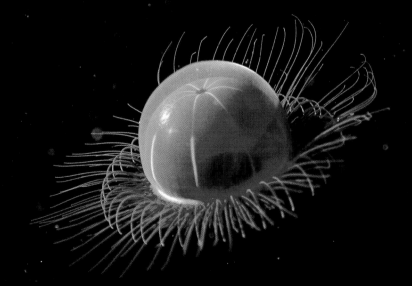

"Under the sea, it seems my every gaze is as stolen from

some forbidden world; and it triggers an emotional shock

that never flags, no matter how many times I dive…"

Jacques-Yves Cousteau, 1976

OPPOSITE
Tiburonia granrojo
The big red
_SIZE 1 m diameter
DEPTH 1500 m

This large dark and velvety ball
was discovered in 1993 by MBARI
researchers in California. It is so
different from other jellies
that biologists had to create
a new subfamily for it, called
the *Tiburoniinae*, after the *Tiburon*,
the robot that discovered them.
To capture its prey, it does not use
stinging tentacles, as do the
majority of jellies; rather it deploys
long fleshy arms whose number
varies, curiously, between four
and seven. Very little is currently
known about this creature.

one for the Great Barrier Reef, none of
the species involved here are found in any
of the warm waters of the tropical seas.

In fact, these natural marvels grow
only in very cold waters, those found at
the extreme latitudes of the planet. Known
since the nineteenth century from the trawls
brought up by oceanographic ships, cold
water coral reefs are also found in the ocean
depths, where no light penetrates—and
where no witnesses can tell about the ravages
they suffer. Shallow water resources are
now diminished to such a point that
the fishing industry is now exploiting
the resources of the deep. Cold water reefs
are found between 40 m and 2000 m
depth. They shelter numerous species of
fishes, crustaceans, and mollusks having
high commercial value. These animals are

Down at the bottoms of the oceans, there
are forests of corals extending over
hundreds of square kilometers, sheltering
an infinitely rich and varied fauna. Sharks
and cephalopods lay their eggs there;
giant gorgonians offer their branches as
promontories for echinoderms; delicate
sponges welcome crustaceans and fishes.
Though the description sounds much like

Deepwater coral reefs:
Out of sight, out of mind

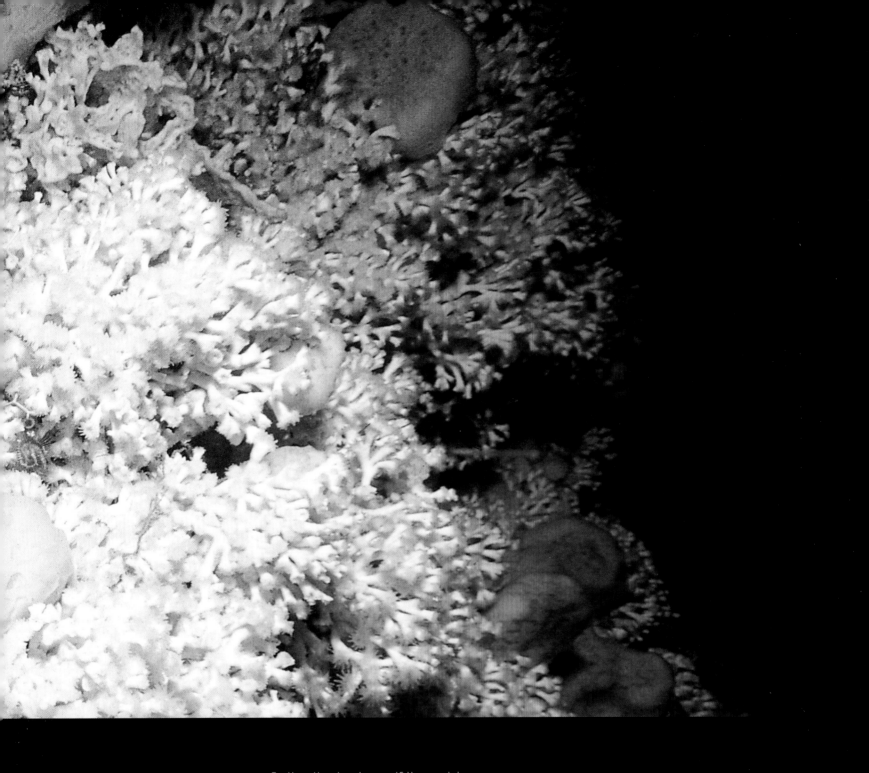

harvested using giant nets, ballasted with weights that drag furrows into the bottom, leaving nothing behind but a landscape of desolation and debris in places where, just a few hours earlier, a scene 8000 to 10,000 years old spread about, with hundreds of species unknown to science and, in large part, endemic to the given habitat. Researchers estimate that in just the last few decades, deepwater reefs were destroyed over a cumulative area several times the size of Europe. The situation is even more problematic because the growth rate of cold water corals is between ten and twenty times as slow as that of the well-known corals of tropical waters.

On the other hand, even if there exist as many coral species in the cold and dark waters as do in the warm waters, the fact remains that only six deep-sea species are capable of constructing reefs. All the other cold corals are "soft" species, incapable of producing calcareous skeletons. These factors make the deep reefs extremely vulnerable to human intervention, but their isolation off the coasts, often hundreds of meters below the surface, makes them ideal victims: out of view and unable to protest or blame. A simple statistic shows how very irrational is the industry willing to destroy such a unique and marvelous natural heritage: the fish caught in these slaughterous underwater raids amounts to only 0.2% of the world's overall marketable catch. •

ABOVE

Gorgonocephalus caputmedusae

Gorgon's head, or basket star

_SIZE central disk 6.5 cm
DEPTH 50 to at least 300 m

Coral reefs extending along the continental margins of Norway are dominated by the tuft coral, *Lophelia pertusa*. Like trees in a tropical forest, they offer a network of hiding places, numerous possibilities for meals and fruitful encounters or associations. This ecosystem attracts a multitude of species, of which the majority may still not be identified by science. This photo has a *Gorgonocephalus caputmedusae* in the center.

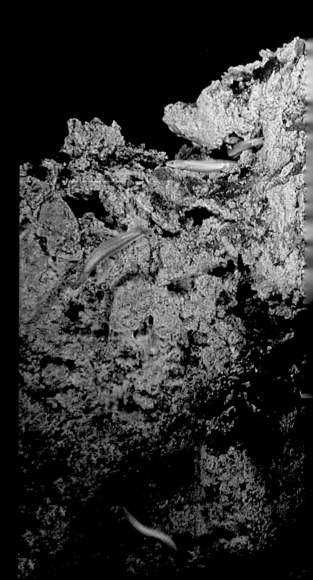

PRECEDING DOUBLE PAGE SPREAD, LEFT

Gorgonocephalus caputmedusae

Gorgon's head, or basket star

_SIZE 1 m diameter
DEPTH 50 to at least 300 m

When fishing, the basket star often climbs onto a gorgonian fan, here a *Paramuricea placomus*, and deploys its arms into the current from there; these limbs resemble the hairdo of its mythological namesake.

PRECEDING DOUBLE PAGE SPREAD, RIGHT

Grimpoteuthis sp.

Dumbo octopus

_SIZE 20 cm
DEPTH 300–5000 m

This little octopus almost seems like a character out of a Japanese cartoon. Researchers have already described fourteen species of *Grimpoteuthis*, but beyond the taxonomic description made on the basis of animals captured by trawlers, these octopuses for the most part are still enigmatic. They are often observed resting on the bottom, with their mantle spread around them. What are they doing there, sitting so quietly in the dark? Nobody knows.

RIGHT

Dysommina rugosa

Cutthroat eels

_SIZE 38 cm
DEPTH 260–775 m

Frightened by the approach of the submersible, these eels leave their crevasses at the top of this very recently formed volcanic cone to the east of the Samoa Islands in the South Pacific. Before 1999, the submarine volcano Vailulu'u had never been mapped at all. Its base is at 5000 m depth and its summit at 700 m below the surface; exploration dives have discovered lava pillars carpeted by an orange-yellow microbial layer. The site was nicknamed the Eel City because of its large numbers of cutthroat eels, observed in a natural habitat for the first time.

"They seem connected to hell itself!" exclaimed Jean Francheteau, one of the first French researchers ever to set eyes on the sensational black smokers around hydrothermal sites. At a depth of several thousand meters, these eerie edifices belch super-heated waters that have circulated through the ocean crust, close to magma, becoming laden with various metals and toxic sulfides. Hydrothermal chimneys give the impression of being the pipeworks of an underground factory linking the oceans to the center of the Earth. Their discovery triggered intense excitement throughout the international scientific community.

Dr. Daniel Desbruyères
French Research Institute for Exploitation of the Sea (IFREMER), France

HYDROTHERMAL VENTS

The deep seafloor, where darkness and cold reign, is the planet's largest, and least known, ecosystem. In spite of its zoological interest, highlighted during the great expeditions of the end of the nineteenth century, it has long been considered a desertic environment. Oceanographic expeditions have shown that in the absence of photosynthetic production, the only food resources available come from the surface, primarily in the form of a rain of particles. This is why the abyssal plains are populated by animals that are quite unusual, few in number, and often very small.

At least that's what's been taught to every oceanographer, biologist, and geologist during the first three-quarters of the twentieth century. Thus, in February 1977, when the American submersible *Alvin* dived to a depth of 2500 m over the Galápagos Ridge, researchers were astonished to discover a profusion of life: communities of strange organisms of spectacular sizes and astonishing morphologies clustered around warm springs (at around ten degrees above the ambient temperature of 2ºC). These luxuriant populations contrast radically with the desolate basaltic environment of the ridge. John Corliss, one of the geologists on board the *Alvin* during this historical dive, communicates with the surface crew:

"Debra," he said to his graduate student, "isn't the deep ocean supposed to be like a desert?"

"Yes," she replied, remembering her ecology course.

"Well, there's all these animals down here!" Corliss exclaimed.

This was the very first discovery of hydrothermal vents. The unusual organisms that researchers found were named in terms of what they resemble, and so we have the "giant tube worm," the "dandelion," the "spaghetti worm," and the "giant clam."

In the following years, researchers avidly explored the ridges, leading to the discovery by a Franco-American team of hydrothermal emissions at extremely high temperatures along the East Pacific Rise. Bill Normark and Thierry Juteau were on their second dive in the Rise Expedition at 21ºN; around noon, the pilot informed the surface that they had come upon "a sort of locomotive smokestack that had some strange thing sticking out of it... There are some odd little fish all along the structure—they look almost like bits of intestine!" The "locomotive smokestack" was actually a hollow chimney-like structure produced from metallic sulfides, and the "strange thing sticking out of it" was a hydrothermal plume, black and dense. The "bits of intestine" were the zoarcid fish (or eelpout), with their crinkly pink flesh. Their habitat, veritable organic forests, is composed of the tangled white tubes of giant worms with red plumes. Fields of clams cover the substratum at the chimney bases, and in the rare vacant spaces, teeming masses of shrimps and crabs take their places. Pompeii worms, so nicknamed by geologists because they live under a constant rain of ashes, secrete tubes that carpet the walls of the hydrothermal chimneys, which can rise to 15 or 20 m in height. Some of the phenomena initially baffled the researchers—but not for long.

Hydrothermal circulation starts off within the crevasses produced by the cooling of magma. Seawater seeps into them, going down as far as several hundred meters in depth, and reacts with the hot rock at temperatures over 350ºC. The burning hot fluid that rises up again is anoxic, acidic, and laden with sulfides, methane, and carbon dioxide, substances that seawater ordinarily bears only in trace concentrations. When this fluid erupts through the fissures in the basalt, the mineral substances precipitate and form chimneys called "black smokers." The surrounding seawater heats up to about 20ºC, making these sites into mild oases—at least in terms of temperature—but the toxicity of the environment is much less inviting. How can such a dense fauna prosper in an environment characterized by toxicity, crushing pressures, and a total absence of light? Discovering hydrothermal vents was like dropping a bomb in the scientific world, producing an explosion of renewed interest for the deep oceans. The scientific explanation of the mechanism permitting life to develop under these extreme conditions was not long in coming.

Bacteria use the chemical substances belched forth by the chimneys to synthesize organic matter, and this serves as the basis for the entire hydrothermal food chain. Stated quite simply, bacteria substitute for green plants in this lightless world, and chemistry replaces solar energy. This process of primary production, whose discovery caused a revolution in biology, is called "chemosynthesis," a term that has already found its place in biology texts, alongside the well-known photosynthesis.

Since the 1970s, hydrothermal vents have been found all along the Pacific Rise, as well as on underwater volcanoes and in basins farther west of the volcanic arcs of the Pacific Ring of Fire. Other sites have been discovered along the Mid-Atlantic Ridge and in the Indian Ocean. All told, more than six hundred new animal species have been described from these astonishing habitats, as well as numerous strains of microorganisms, some of which live at extremely high temperatures and secrete rare compounds that have biotechnological applications.

The discovery of chemosynthesis in the deep ocean was possibly the most striking scientific discovery in the field of oceanography during the twentieth century, so much so that we now think in terms of "before 1977" and "after 1977." Other chemosynthetic habitats have since been identified, like cold methane seeps or whale carcasses, but in both of these cases chemosynthesis is connected to primary production that occurred at the surface. The methane and other hydrocarbons at the basis of populations around cold seeps are examples of fossil energies, and as the name indicates, these were formed through the accumulation of organic matter on the bottom of the oceans, with the passage of millions upon millions of years. Similarly, whale carcasses are a direct product of the system of photosynthesis that reigns at the surface of the globe. Thus, the ecosystems of hydrothermal vents depend less upon the sun's energy than any other on our planet. Their discovery has not only challenged our perceptions of the deep oceans, which now reveal themselves, under certain conditions, to harbor an extraordinary biological wealth, but it has also very rapidly raised new questions about the origins of life on Earth. At the dawn of our planet, the hydrothermal oceanic habitat was doubtless much better represented than it is now; it is even quite probable that the first molecules of life were synthesized there. This scenario is proposed today as an argument for searching for life on other celestial bodies, like Europa, one of Jupiter's moons. The idea of extraterrestrial life—but in the form of microorganisms and not of Martians, as we have fantasized about for over a century—is no longer eccentric or improbable. •

Paralvinella palmiformis
Palm worms
_SIZE up to 15 cm silhouette
DEPTH 1530–2700 m

Palm worms live in dense colonies around hydrothermal vents found along the northern section of the Mid-Atlantic Ridge, off the coast of Canada. Though the surrounding waters are laden with metals that would be lethal to humans, they do not seem to inconvenience the worms. They form colonies that resemble impregnable bushes, but they don't seem to repel the gastropods and scale worms seen here dusted with organic and mineral particles.

THE HYDROTHERMAL PLUMES ESCAPING FROM THE MIDOCEAN RIDGES AT

WITH A BURNING

OVER 300ºC REMIND US THAT THE EARTH IS A LIVING PLANET CORE THAT IS COOLING LITTLE BY LITTLE.

In certain areas along the midocean ridges, tectonic plates can separate at up to 18 cm per year, forming valleys hundreds of meters wide, which are called rifts. The magma that gathers there to fill the vacant space cools abruptly when it makes contact with the glacial waters of the abyss. This then ends up caving in and fashioning a desertic, almost lunar landscape, scattered with chimneys spewing thick, toxic fumes. Explorers who have witnessed the raw forces of nature at work here have all felt like they had traveled to the primordial cradle of the Earth, at the roots of the formation of the universe.
The two Russian submersibles appearing in this computer-generated image, *Mir I* and *Mir II*, are among the very few vessels capable of exploring the entire ridge system.

RIGHT
Riftia pachyptila
Giant tube worm

_SIZE up to 2 m
DEPTH 2000–2850 m

Up until 1979, whenever one thought about worms, the image of a colorless earthworm came immediately to mind, but the discovery of the giant, sublimely colored creatures living around hydrothermal vents in the eastern Pacific abruptly changed that view. These astonishing creatures live in symbiosis with the chemosynthetic bacteria that provide the worms with their meals. It took the specialists a while to understand the functioning of the animal, which at first they believed to be a filter feeder. Robert D. Ballard remembers their incredulity: "With no eyes, no mouth, or any other obvious organs for ingesting food or secreting waste, and no means of locomotion, it was no worm, snake, or eel, but no plant either— the strangest creature we had ever seen."

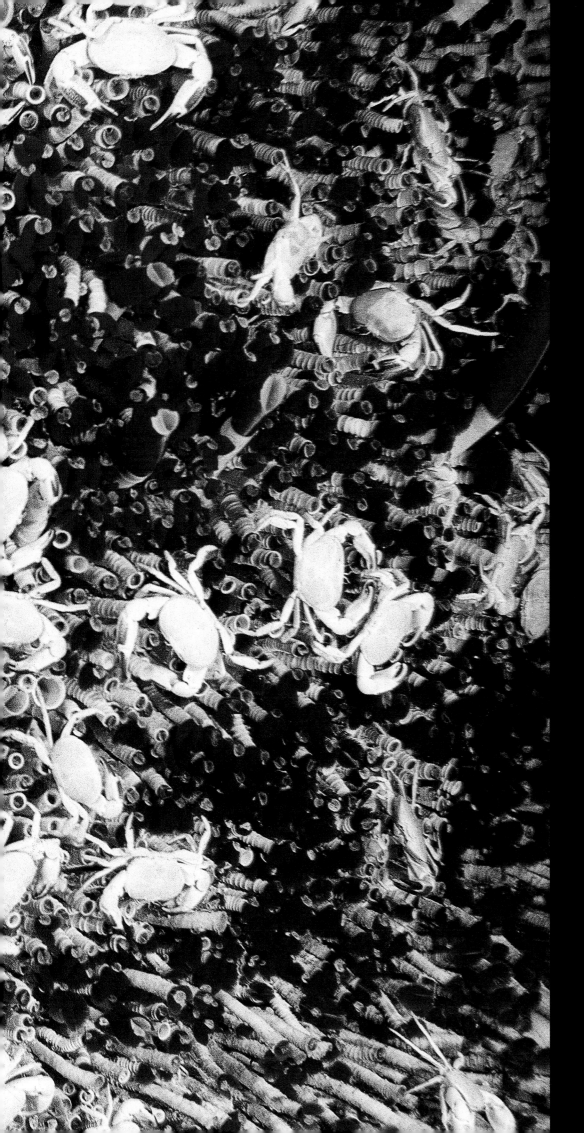

Tevnia jerichonana

Jericho worm

_SIZE 35 cm
DEPTH 2600–2850 m

There is probably a lot of territorial competition among hydrothermal creatures trying to gain access to the sites best exposed to the circulation of fluids. This portion of the bush has visibly been overtaken by Jericho worms, which are smaller than their close neighbors, the *Riftia* giants, a few of whose tubes can be seen coming out in the photo. Like the *Riftia*, they harbor bacteria that transform toxic substances into food for their hosts, and accordingly, choice of location is of paramount importance for these sessile animals. When they are threatened by approaching *Bythograea thermydron* crabs, which cut off portions of their gills, they retract into their tubes.

Alvinella pompejana

Pompei worm

_SIZE 15 cm
DEPTH 2600–2850 m

The Pompei worm is the only creature on Earth that bathes daily in temperatures around 80°C. It lives on the walls of active chimneys that regularly collapse under their own weight, destroying the network of organic tubes secreted by the worm. It remains a mystery how this creature is able to resist the extreme toxicity of its habitat. Perhaps the white bacterial filaments that look like hairs are used to neutralize the chemical pollution.

Kiwa hirsuta

Yeti crab

_SIZE 20 cm
DEPTH 2300 m

Whether it's King Kong or a Yeti, this crustacean with its long hairy arms definitely evokes an image of some legendary primate. Discovered in 2005 along the ridge running at more than 2000 m depth to the south of Easter Island in the Pacific, this little albino lobster, hairy and blind, provoked a media tidal wave. Michel Ségonzac, who discovered it, asked: "Why this sudden keen interest? Are people trying to escape the present through some cuddly-toy sentimentality?" *Kiwa hirsuta* had an immediate effect in terms of affective projection; in the very week following the announcement of its discovery, a stuffed animal in its image was produced in Japan!

The hydrothermal biomass is between 10,000 and 100,000 times as great as that of the general abyssal environment.

Characteristic of this profusion are the fields of *Bathymodiolus septemdierum* mussels (12 cm long, LEFT PAGE, BOTTOM) and the teeming colonies of blind *Rimicaris exoculata* shrimp (5 cm long, RIGHT PAGE, LOWER RIGHT), in which up to 2,500 individuals can be found per square meter.

Not only the creatures living in symbiosis with chemosynthetic bacteria are able to attain such high densities. Even "normal" animals form aggregations of proportions found nowhere else: the *Freyella* brisingid echinoderms (*Freyella* sp., 50 cm diameter, RIGHT PAGE, TOP LEFT) or the predatory anemones *Pacmanactis hashimotoi* (4 cm high, LEFT PAGE, TOP) are generally solitary animals found only intermittently in the abyssal plain.

The vent stauromedusa or stalked jellyfish (*Lucernaria janetae*, 12 cm high, CENTER) belongs to a genus previously known only at the surface; it has been discovered, though, in phenomenal concentrations around a hydrothermal vent at 2750 m in the eastern Pacific!

Just like an oasis in the desert, these toxic sites attract an entire peripheral fauna constituted of sessile crustaceans, such as the *Vulcanolepas "Lau A"* sp. vent barnacle, which resembles a flower (RIGHT PAGE, MIDDLE RIGHT), or mobile crustaceans, like the spider crab (*Macroregonia macrochira*, over 50 cm, RIGHT PAGE, LOWER LEFT) seen here carrying its eggs.

The little depigmented vent octopus (*Vulcanoctopus hydrothermalis*, 20 cm, LEFT PAGE, MIDDLE) can often be observed striding along the giant tubes of *Riftia* worms.

Lamellibrachia sp.

Asphalt worms

_SIZE 70 cm
DEPTH 2900 m

For about forty years, we have known that the overwhelming majority of the Earth's volcanic activity occurs in the oceans. Even so, the discovery that lava can be made up of pure asphalt, as is the case for that covering the hills south of the Gulf of Mexico at 2900 m depth, was a surprise. Furthermore, the fact that living creatures can extract the energy needed for life and growth from this asphalt is simply astonishing. As researcher Ian McDonald says, "It shows how this planet is alive from top to bottom, and how it always maintains a capacity to surprise us."

Dr. Lisa Levin
Scripps Institution of
Oceanography, USA

gas promotes mass: methane seeps

Imagine a Jacuzzi with jets set on low, or your soft drink after a good shaking. Now pretend those bubbles are filled with methane rather than air or carbon dioxide. There you have the scene at some of the strangest, and most recently discovered, ecosystems on the seafloor: methane seeps. Here aggregations of clams, mussels, and tubeworms thrive on the chemicals emanating from the seabed. Prior to the discovery of seeps, scientists thought that chemically driven systems on the deep seafloor were associated only with hot vents. Seeps proved them wrong. First revealed in 1984, cold seeps have since been discovered throughout the world's oceans.

Methane seeps occur from the shallow subtidal zone to the ocean trenches, at depths ranging from 15 m to more than 7800 m. Thus, they are not exclusively a deep-sea phenomenon; however, only those systems below the continental shelf ($\geq$ 200 m) host highly specialized biological communities.

Methane is a clear, highly combustible, odorless gas, familiar to all as a source of energy for our gas stoves and home heating. Natural gas, recovered by drilling, is about 75% methane. It smells only because organic sulfur compounds are added so that gas leaks can be detected. Methane is found in the Earth's crust under the ocean. In areas of high primary production, large amounts of organic matter—mainly plankton—are deposited over millions of years in the seabed along the edge of the continents. As the organic material sinks and accumulates on the seafloor, it becomes buried under layers of sediment. Then microbes (or the effect of pressure and heat in certain areas) decompose the organic matter without any oxygen, resulting in the formation of methane.

When the deep-buried methane moves upward towards the seafloor, it is consumed by microbes that interact with other bacteria to produce sulfide. Although sulfide, which smells like rotten eggs, is usually highly toxic, it supports a suite of animals that are specialized in dealing with chemical environments; these are chemosynthetic animals, similar in their body organization to the animals found at hydrothermal vents. The fauna attracted to the methane seepage form true animal oases on a landscape of otherwise relatively featureless, homogeneous sediment in the deep sea. First, bacteria graze on the chemical fluids (methane and sulfide) that seep out of the seafloor. Then, special clams and mussels arrive that house symbiotic bacteria that can harvest the chemicals to produce energy for their hosts. Also present at seeps are tubeworms with sulfide-consuming bacteria; some have very long roots that can reach a meter down under the seafloor to look for sulfide.

Methane formed within the seafloor can be squeezed upward by the subduction of oceanic plates under the continental margins. That's why these new ecosystems were named "methane seeps" when they were first discovered. Since then, scientists have learned that methane does not always "seep" out of the seafloor. It can also be exposed by earthquake-induced landslides. For these reasons methane seeps are common along the entire Pacific rim: off Japan, Alaska, Oregon, California, Costa Rica, Peru, and Chile, all areas where there is a great deal of tectonic activity. Seep communities can also occur in other settings, in association with hydrocarbons such as petroleum oil, tar, or asphalt.

The first animal communities ever seen living at seeps on the deep seafloor were found in 1984 off Florida in the deep Gulf of Mexico, associated with brines containing sulfide. Brine is a very salty water that oozes out of large salt deposits deep within the Earth's crust. Two hundred million years ago, the Gulf of Mexico became an isolated sea, which dried out entirely, producing an 8 km–thick layer of salt. Later, a passage linked the gulf to the oceans again. Today, the salt layer is trapped under millions of years of sedimentation, but plate movements called "salt tectonics" cause this buried substance to reach the seafloor, where it sometimes forms distinct "brine lakes."

The global inventory of methane in the deep ocean may be ten times that of the conventional oil reservoir, gas fields, and coal beds combined. New seeps are discovered every few months and will probably prove to be much more widespread than hydrothermal vents, whose geographical distribution is essentially linked to the volcanic activity at midocean ridges or behind subduction zones.

At high pressure and low temperatures in the deep sea, methane can occur in solid form, known as gas hydrate. The hydrates form by the movement of methane gas upward in the seabed along faults and cracks. Contact with cold water causes crystallization: methane becomes trapped in a prison of water molecules, forming a solid ice within the seafloor. Massive quantities of methane hydrates occurring along the continental margins could represent a major future energy source, as 1 liter of methane hydrate contains 168 liters of methane gas. However, the solid form is stable only at high pressure and cold temperatures, posing a challenge for recovery, transport, and implementation as a fuel source. A big problem linked to the use of methane remains; it is a greenhouse gas, warming the atmosphere far

more than carbon dioxide. Some theories suggest that massive release of gas hydrates long ago in the Earth's history could have triggered rapid severe warming of the atmosphere. This may have occurred during the Permian-Triassic extinction event 252 million years ago for example, or during the Paleocene-Eocene thermal maximum 55 million years ago, but this idea is still under debate.

Seep systems host a wealth of biodiversity, from microbes to mussels. Strange new microbial interactions and relationships are emerging with every visit to new seeps. The animal communities inhabiting cold seeps are similar to those at hydrothermal vents, but differ in the absence of elevated temperature, and in having greater longevity. Seep emissions occur at temperatures similar to those in surrounding sediments, thus they are sometimes called "cold seeps." Seep fluid emissions, while shifting positions locally, are thought to persist in a particular area for much longer periods of time than many hydrothermal vents (venting is inherently ephemeral, at least where the ridge crest spreads quickly), creating more stable communities with longer-lived organisms. Seep tubeworms, for example, may live for over two hundred years!

The study of methane seeps is still in its infancy. We have yet to discover most seeps and perhaps most seep species. We don't yet know how seep animals reproduce, move between seeps, respond to settlement cues, or interact with one another. Better understanding of seep ecosystems may ultimately unlock secrets about climate change, the evolution and maintenance of life in the deep sea, and possibly even life on other planets, where oxygen is scarce and toxic chemicals abound. •

OPPOSITE
Lamellibrachia luymesi
Cold seep tubeworm
_SIZE 2 m
DEPTH 1000 m

The life span of this tubeworm is among the longest in the animal kingdom: 250 years! Like its close hydrothermal cousins, it "eats" hydrogen sulfide via the bacteria it cohabits with. Unlike its relatives, the cold seep tubeworm has roots that it drives into the substrate, seeking essential resources. It is also possible that it uses the roots to inject its sulfate excrements back into the sediment. This ingenious method would permit it to stimulate beneath its very feet the hydrogen sulfide production that nourishes it and thus explain the creature's exceptional longevity. The tubeworm lives in the vicinity of hydrocarbon seeps, which are much more stable habitats in the long run than hydrothermal vents.

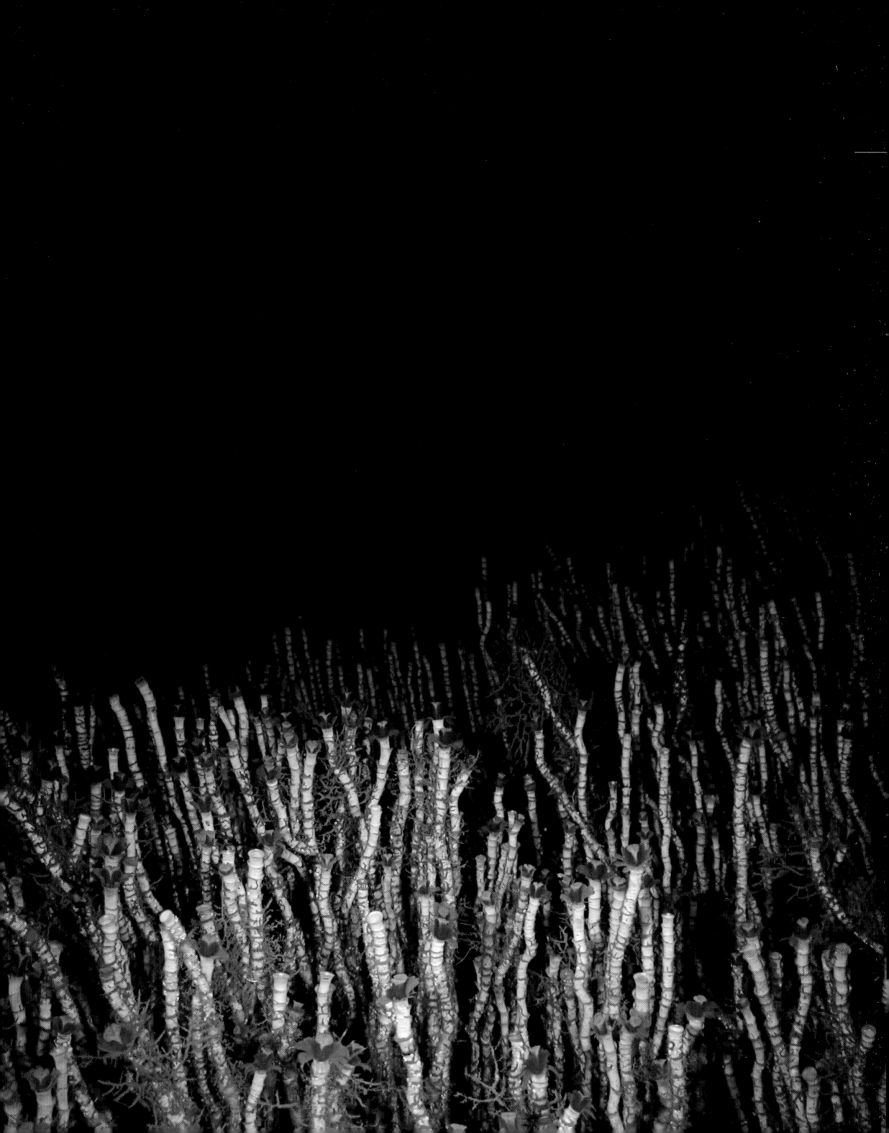

"By the late 1960s, the deep seafloor had completely reversed

its image, from a rather boring geological dead zone

to the hotspot of current research."

Robert D. Ballard, 2000

OPPOSITE TOP

Hesiocaeca methanicola

Iceworms

_SIZE up to 5 cm
DEPTH 540 m

Which of the two ought to be
considered more unusual? The pink
worm working nonchalantly in a ball
of methane-flavored ice, or the
ice itself? When temperatures are
low and pressures high, methane
crystallizes within a prison of water
molecules, forming little heaps of
ice called methane hydrates.
This phenomenon is a curiosity in
itself, but the discovery in 1997
of polychaete worms sculpting the
orange hydrate surfaces reminds us
that we have not yet been introduced
to all the strange phenomena of
our planet.

OPPOSITE BOTTOM

Methane hydrates are generally
white, but here they are soaked
with hydrocarbons that color them
orange. The gas bubbles that
attempt to escape from their ice cage
are made heavier by a viscous
coating of oil, giving them
the appearance of mushrooms.

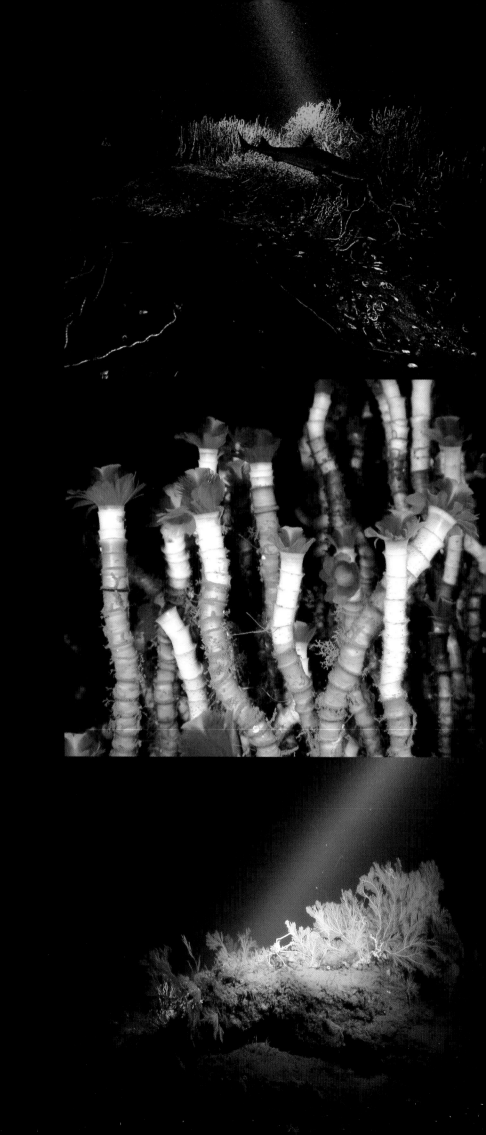

What is particularly admirable about chemosynthetic ecosystems, whether they are linked to emissions of hot fluids along the ridges or to gas or hydrocarbon seeps along the continents, is the sudden blossoming of a formidable food chain within a context that otherwise can be considered pretty dismal.

The oasis relies on the presence of bacteria that transform the gases into hydrogen sulfide. Next come the creatures living in symbiosis with the bacteria feeding on this sulfide, like the cold seep tubeworms (*Lamellibrachia luymesi*, LEFT PAGE, MIDDLE) or the *Escarpia spicata*, seen here together with *Lamellibrachia barhami* (CENTER).

Other animals, like snails, graze freely on the bacteria covering the surfaces of the methane hydrates (RIGHT PAGE, MIDDLE RIGHT). One can clearly see around this heap of ice that the gas can sometimes escape from underground layers in columns of bubbles, as though the ocean floor were leaking.

A nonspecialized peripheral fauna rapidly springs up because of the site's rich biological production; the elegant painted squat lobsters (*Eumunida picta*, RIGHT PAGE, UPPER LEFT) are typical denizens of the deepwater coral reefs. They have been observed capturing fish that swim within reach. The roughskin spurdog (*Cirrhigaleus asper*, LEFT PAGE, TOP) loitering idly around among the worms' tubes and the conger eel (*Conger* sp., RIGHT PAGE, UPPER RIGHT) are generalist animals that come to hunt in favorable terrain.

Like the squat lobsters, the gorgonians (*Callogorgia americana*, RIGHT PAGE, LOWER RIGHT) frequently occupy deep coral reefs. An entire field of these animals was recently found covered with shark eggs. Here, it is the brittle star (*Asteroschema* sp.) we see decorating the gorgonians' branches.

OPPOSITE TOP
What a bleak, uninviting landscape
is presented by the layer of fog
surrounded by an expanse of
mussels seeming to float
weightlessly. Of all the chemical
phenomena unfolding in the deep,
the brine lakes are one of the most
bizarre. The very existence of
submarine lakes within the ocean
defies comprehension. How is it that
their waters avoid mixing with the
surrounding waters? Quite simply,
because of the extremely high
salinity. The brine comes from layers
of salt buried hundreds of meters
beneath the ocean crust. Like
another Dead Sea transplanted to
the depths of the Gulf of Mexico,
this lake has such a high salinity
that exploration submersibles
bounce upward when they touch its
surface; they float and can even cut
their engines, providing "a very
surreal experience," in the words of
Charles Fisher.

OPPOSITE BOTTOM
The brine is often enriched with
methane through the course of
its rise toward the ocean floor.
It is thus exploited by animals
adapted to this sort of alimentary
regime, like the cold seep mussels
(*Bathymodiolus childressi*) carpeting
the shores of this underwater lake.
Nonetheless, the line between life
and death is very fine, because
the salt concentration is fatal to
all creatures. Living beings need
to stay on the outskirts of this
cemetery, which preserves, floating
in its brine, the skeletons of the

"The endless frontier of the immediate future

is not outer space but the vast space beneath the sea."

Robert S. Dietz, 1961

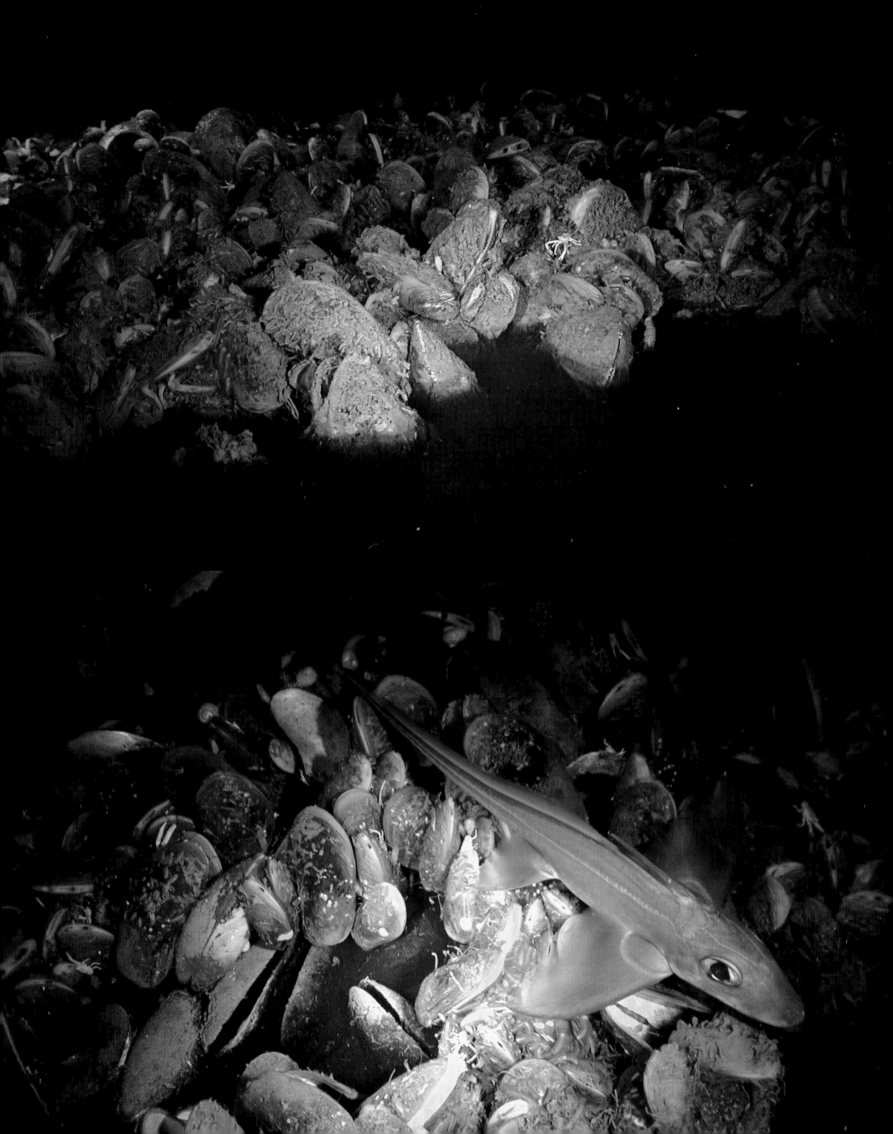

Scientists have long known that most whales, upon death, sink into the deep ocean—not in a mythical whale cemetery, but rather along migration routes where physical stresses are high. These giant creatures are the largest animals on Earth—an adult blue whale reaches 30 m in length and weighs 100 tons—so scientists reasoned that the arrival of a dead whale at the normally food-poor deep seafloor must create a huge bonanza for scavengers. However, nobody even dreamed of the rich animal communities that live on whale remains until 1987, when the first whale carcass was directly observed in the deep ocean.

Then, the research submersible *Alvin* chanced upon a whale carcass at the seafloor; at about the same time, the U.S. Navy found eight whale skeletons while searching for a crashed missile off the California coast. This spurred inquisitive scientists to experimentally sink and study at least twenty more whales that had died from natural causes. These studies reveal that dead whales support an unexpected diversity of life, and that hunting of whales in the surface ocean could endanger numerous species living off whale carcasses in inky darkness kilometers below.

The deep seafloor is normally a very food-poor place fed by a weak rain of tiny organic particles sinking from the surface ocean. A sunken whale is thus a massive organic input to the underlying seafloor, providing in a single pulse as much food as would normally arrive in four thousand years. The consumption of a large whale carcass at the deep seafloor can take many decades, perhaps even a century. The dead whale, or "whale fall," is utilized by a predictable succession of animal species that exploit the blubber, muscle, and bone in many ways. When

Dr Craig R. Smith
University of Hawaii, USA

a dead whale first hits the bottom, its shock wave and foul odor plume can attract hundreds of highly mobile scavengers within hours. Off California, the main scavengers include 4 m–long sleeper sharks that slice massive chunks from the carcass, as well as hundreds of slimy, eel-like hagfish that burrow into the blubber as they feed. All told, there are at least thirty-eight species of fishes, crustaceans, and mollusks that form a deep-sea feeding frenzy on the whale's soft tissue, removing up to 60 kg per day from the carcass; this frenzy lasts for months on the remains of a juvenile gray whale and probably more than a decade on an adult blue whale, ultimately reducing the carcass to a skeleton.

Not surprisingly, the frenzied scavengers are sloppy feeders, dropping bits of whale flesh onto the seafloor all around the carcass. This sloppy feeding causes organic enrichment of sediments near the whale fall, allowing a suite of worms, snails, and crustaceans to thrive in this mixture of mud, rotting flesh, and bacteria. These animals are called "enrichment opportunists" because they are attracted from long distances (at least tens of kilometers) to the organic-rich conditions next to dead whales; they seize this opportunity to grow and breed rapidly, broadcasting their young into the water column

A WHALE'S END IS
THE BEGINNING OF LIFE
AT THE DEEP SEAFLOOR

to drift on ocean currents to the next whale fall or other enrichment event. Quite surprisingly, a number of the enrichment-loving worms and snails are new to science and have been found only around whale falls; they may have evolved beneath whale migration routes, where dead whales commonly sink to the seafloor to form persistent, organic-rich habitat islands. The most bizarre animals in this assemblage may be the bone-eating worms, sometimes called snot worms because they exude huge amounts of mucus, that show amazing adaptations for consuming whale bones and the rich whale oil the bones contain. These worms look like tiny red palm trees, with a green root system that excavates cavities within the bones. The worms' "roots" also contain a special bacterial garden that can degrade the whale oil, providing a rich food source to the worm. This is the first animal known to actually break down large mammal bones at the seafloor to mine the rich food material trapped within the bone matrix.

After a number of years the enrichment community wanes, and a new "sulfur-loving" group of animals takes over the whale skeleton. This community thrives on chemical energy in sulfide that oozes from the bones. The sulfide is produced by bacteria, which use anaerobic metabolism to degrade the tons of whale oil still contained within a large skeleton. The degradation of whale oil is initiated by oxygen-breathing bacteria; however, the enormous reservoir of oil in the skeleton quickly exhausts all available oxygen, causing a switch to sulfate-breathing bacteria that transform sulfate from seawater into sulfide, an energy-rich compound. The production of sulfide leads to a whole new chemically driven food chain around the carcass. Giant clams and tiny mussels now colonize the bones in the thousands, each with an internal colony of bacteria that can "eat" sulfide and then use the energy from this highly toxic compound to produce organic material to feed their host mollusk. These clams and mussels are closely related to some of the exotic species first discovered at deep-sea hydrothermal vents, and there is intriguing evidence that sulfide-rich whale falls have provided stepping stones for the colonization of vent habitats over the last thirty million years. Along with clams and mussels, a complex food chain of more than four hundred other species of animals share the sulfide-rich whale skeletons. Many of these animals have been found only living on whalebones and may be "whale-fall specialists"—they must find a whale skeleton on the seafloor to complete their life cycles. This may not be as crazy as it seems: it is now known that a large whale skeleton can support a sulfur-loving community for many decades, and that there are likely to be six hundred thousand active whale falls at the bottom of the world's oceans at any given time!

The possibility that a substantial number of deep-sea animals have evolved a dependence on whale falls raises interesting consequences for the whaling activities of humans. In many oceans, the hunting of great whales has reduced the number of living whales by as much as 75%, which in turn has reduced the number of whales naturally dying and sinking to the seafloor. This means that whaling has caused a large reduction in the amount of habitat available to whale-fall specialists. It is well known from ecological studies that dramatic habitat loss inevitably drives many species to extinction. Thus, we can expect that if the great whale populations are kept low by continued whaling, humans may extinguish numerous deep-sea species that are vainly waiting for the once abundant rain of dead whales from above. An important lesson here is that ocean ecosystems are connected in ways we never dreamed of but continue to discover; the deaths of leviathans in the surface ocean support a remarkable diversity of life thousands of meters below at the deep seafloor. •

OPPOSITE
Cetaceans are unique among mammals: their bones are made up of 60% fat, which augments the buoyancy of their gigantic skeletons within the water. After the first stage of decomposition, which is dominated by the large, mobile scavengers, and the second, characterized by the small fauna that take responsibility for the fine cleaning, the third phase is marked by the work of bacteria. These penetrate to the interior of the bones to decompose the lipids in them, liberating the hydrogen sulfide that will serve as a basis for an entire chemosynthetic food chain. A veil of bacteria covers over the whale's vertebrae like a layer of powdery snow, waiting to be grazed upon and consumed by myriad creatures.

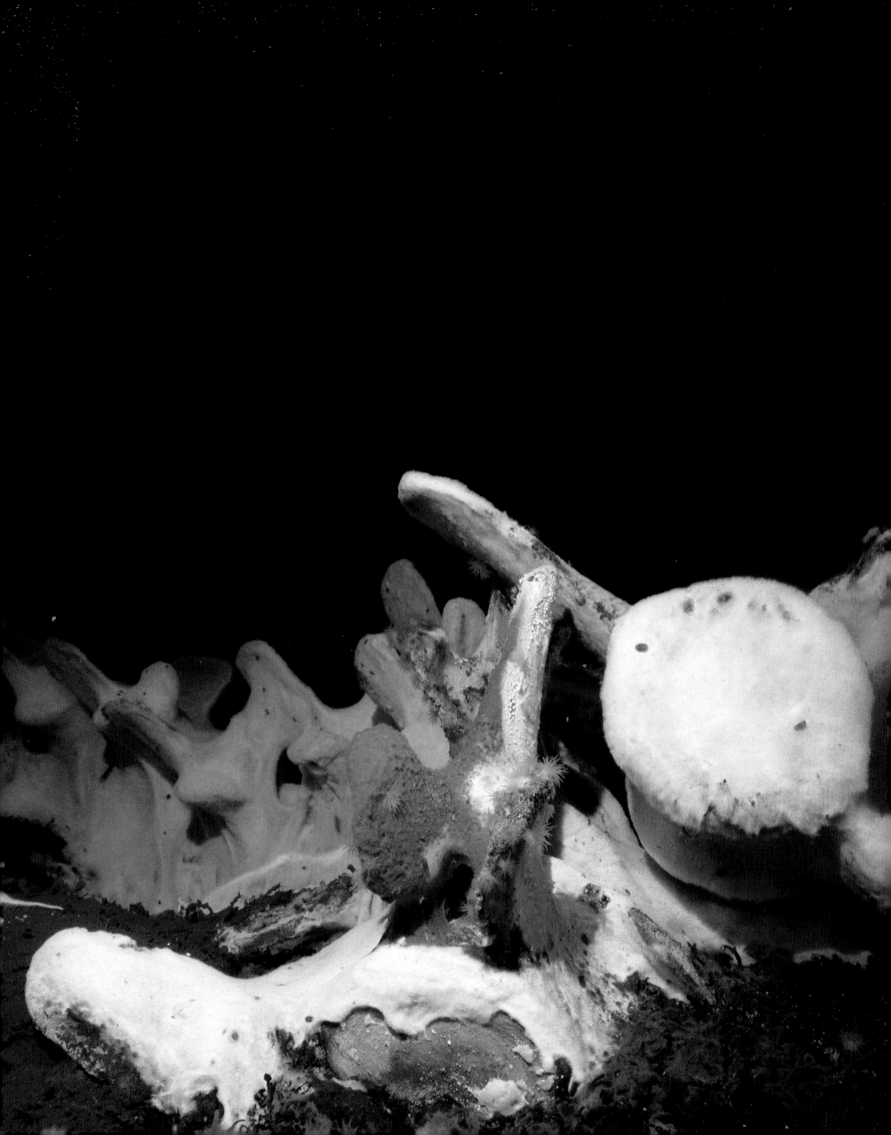

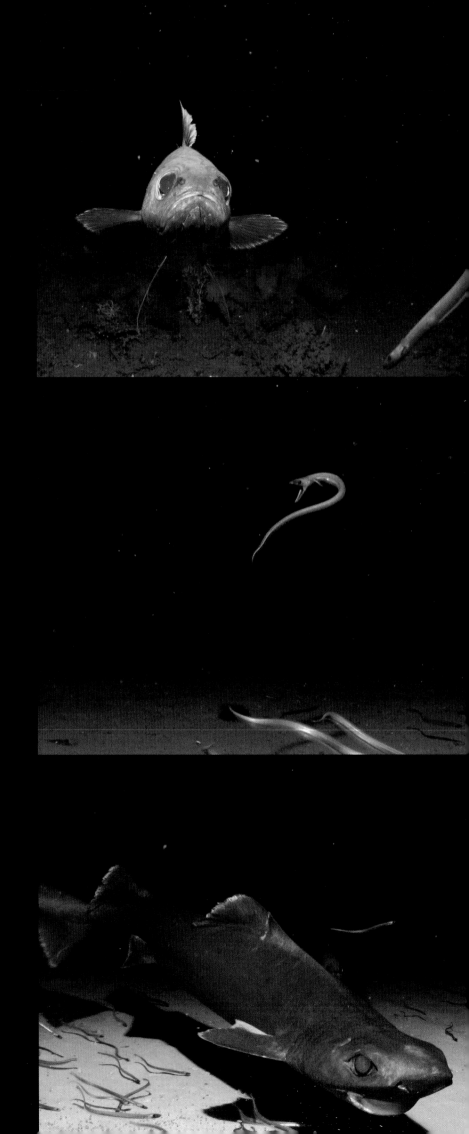

TOP AND OPPOSITE
Mora moro
Common mora
_SIZE up to 80 cm
DEPTH 450–2500 m

CENTER
Synaphobranchus kaupii
Kaup's arrowtooth eel
_SIZE up to 1 m
DEPTH 236–3200 m

BOTTOM
Centrophorus squamosus
Leafscale gulper shark
_SIZE 160 cm maximum
DEPTH 145–2400 m

Being a scavenger of the depths
does not mean one can't hunt a live
meal once in a while. In general,
carrion feeders are opportunists
that seize upon any meal they find.
The common mora, for example,
a commercially exploited fish,
not only consumes dead whale flesh,
but also living crustaceans and
matter discarded by humans.
The Kaup's arrowtooth eel is not
meaty enough to interest the fishing
industry, unlike the leafscale
gulper shark, which is one of
the most exploited species living
on the continental slope.

The deep domain of ocean trenches below 6000 m is named "the hadal zone" after Hades, ruler of the underworld in Greek mythology. Trenches are so remote from our everyday experience that they have bewildered people's imagination for many years. At up to 11000 m below the surface, they represent just over 1% of the seafloor, but how radically different from any other ecosystem in the oceans! In the deeper trenches, crushing pressures exceed a ton per square centimeter. One might be skeptical that life could persist under such extreme conditions. But it does. How much do we know about these deep gaps that cut into the Earth's flesh like the marks left by some giant claws?

When you imagine the deepest point in the ocean, your mind takes you far from shore, but the reality differs: the center of the ocean is often relatively shallow and the deepest trenches in the world are close to land. More than thirty deep trenches exceed 6000 m in depth, with four of them reaching almost 11000 m. Even if you placed Mount Everest (8848 m) upside down into one of these trenches, it would leave an extra 2000 m. Three quarters of the Earth's trenches are located in the Pacific Ocean. They are the vertiginous expression of the violent forces at work in the rocks beneath our feet, or perhaps more accurately, beneath our boats.

The Earth's surface is made of several plates, rather rigid boards that float on a viscous, underlying liquefied rock. Where plates drift apart, magma wells up to fill the gap and creates huge underwater mountain ranges called midocean ridges. As new rock forms, older rock is pushed aside until one plate slides under the other, forming a trench. In trenches, old, heavier plates sink down under young, buoyant plates.

By 1912, most of the world's trenches had already been mapped but the deepest point in the world was yet to be found. Many nations raced to locate it, which a British vessel did in 1950. That point in the Pacific, the Challenger Deep, lay 10924 m deep in the southern Mariana Trench.

On January 23, 1960, U.S. Navy lieutenant Don Walsh and Jacques Piccard—son of the brilliant Swiss scientist Auguste Piccard, inventor of the bathyscaphe—achieved the impossible: in a bathyscaphe called the *Trieste* they dived to a depth of 10916 meters in the Challenger Deep. As they touched the bottom after a five-hour descent, Piccard and Walsh observed a flat fishlike animal that slowly moved away from the sediment raised by the bathyscaphe. Although fishes have never again been reported from these depths, this proved that life was able to survive under such extreme pressures.

The *Trieste* is, to this day, the only submersible to have taken men to the deepest point on Earth. Piccard and Walsh's historical dive put a definite end to the frantic race for records. After the era of conquest came an era of true exploration, and most submersibles were therefore designed to reach a maximum depth of 6000 m, which put 97% of the world's oceans within their grasp. Of all countries, Japan had the best reasons to study trenches: it is surrounded by these deep scars and is home to frequent, violent seismic shaking in the form of earthquakes. Japan invested in a unique investigative tool: a deep-diving tethered robot called *Kaiko*, capable of reaching a depth of 11 km. Through many dives to 10000 m or more, scientists in Japan have established not only that life is common under extreme conditions, but that sometimes these conditions are a prerequisite for the survival of animals: the "extremophile" organisms. Some microbes, for example, can fulfill their life cycle successfully only if they live in a highly pressurized environment.

Dr. Kantaro Fujioka
and Dr. Dhugal Lindsay
Japan Agency for Marine-Earth Science
and Technology (JAMSTEC), Japan

Deep Trenches:
The Ultimate Abysses

Instead of putting a halt to the development of life, extreme pressure can be a requirement of good health!

But there are more than just microbes thriving at the bottom of deep trenches. Slow currents carry gelatinous, floating creatures, swept along just above the bottom—sea cucumbers (or holothurians). In the food-scarce world of the deep sea, gelatinous bodies are a common adaptation. Animals can increase their size without investing too much mass and energy into making hard body parts. In the quiet waters of the deep there are no waves to break such jelly-like animals to pieces, so the aqueous bodies of sea cucumbers are perfectly adapted. In fact, below 8000 m, over 98% of a trawl catch is made up of holothurians. Along with sea cucumbers, the remotely operated vehicle (ROV) *Kaiko* discovered scattered polychaete worms and xenophyophores, giant single-celled organisms, 10 to 25 cm across that look like globs of detritus. Although unseen by the cameras, a core sample of sediment revealed masses of microscopic organisms called foraminiferans, some of which molecular studies revealed to be living fossils—the "coelacanths" of trenches. Foraminiferans are unicellular animals and an important link in the food chain in many environments. Finding abundant colonies of these living fossils alive at the deepest point on Earth was overwhelming. Trenches also harbor swarms of carnivorous animals such as scavenging amphipods, small crustaceans with huge guts, which lie under the surface of the ooze waiting for dead fish and other animals to sink down onto the mud and nurture them.

Other weird and wonderful trench inhabitants are the wood-fall specialists encountered there. It is a strange thought that animals living so far from the surface depend on plant matter to survive. Because trenches are so close to land and so steep, the fresh vegetation torn by storms and typhoons sinks to immense depths very quickly. These wood-falls provide an energy source at the bottom, allowing specialized wood-boring worms, clams, and crustaceans to thrive.

Some trench dwellers do not need to rely on the rain of corpses and debris from surface waters to sustain them. Chemosynthetic animals live off the gases and fluids that are squeezed out of the sediment by plate movement. Where noxious liquids rich with methane or sulfides seep out of the seafloor, animals with symbiotic bacteria—clams, mussels, and tubeworms— thrive. *Kaiko* has discovered such chemically driven communities as deep as 7300 m.

A unique trait that characterizes trench biological communities is the high level of endemism. Trenches are steep and run parallel to each other, so they act as isolated deep valleys— making animal exchanges virtually impossible. Thus, animals become genetically isolated and new species evolve. Up to 50% of sampled animals prove endemic to a single trench.

The variety of trench organisms teaches us that life can survive even the most challenging environments. These deep chasms almost certainly harbor many more organisms new to science. To date, the midwater below 6000 m has never been explored, with efforts focusing on the deep seafloor and the animals living above and within it, while ignoring the huge volumes of water lying above. The challenges are greater, of course. Imagine trawling a plankton net with over 15 km of cable! The accidental loss of *Kaiko*, the world's only active trench explorer, in 2003 means that the quest for knowledge on trench inhabitants has been halted. It was a vivid reminder of the immense technological challenges facing those who wish to know more about our planet's deepest secrets. •

NEXT DOUBLE PAGE SPREAD
Unidentified species
Family Eusiridae
_SIZE 2 cm
DEPTH 2600 m

With its curved spines and prehistoric appearance, this deep-sea amphipod numbers among the most striking species of its genus. It lives on the ocean floor, where it clings to the stalks of glass sponges using its hooklike legs. Amphipods are widespread and among the deepest-living animals in existence, having been found in ocean trenches all over the world. They are voracious scavengers that descend in avid swarms on dead animals that have sunk from above. The head can be seen to the right of the photo; the bright pink parts are its jaws and mouth.

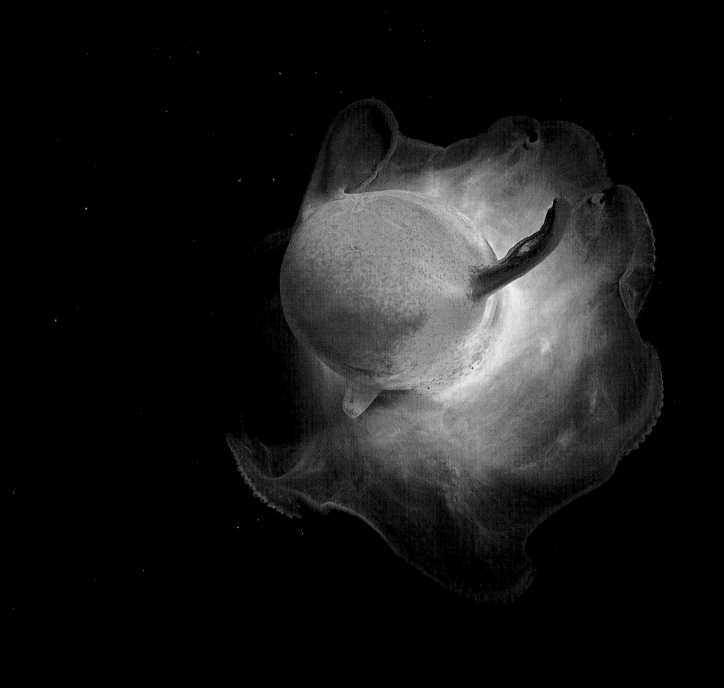

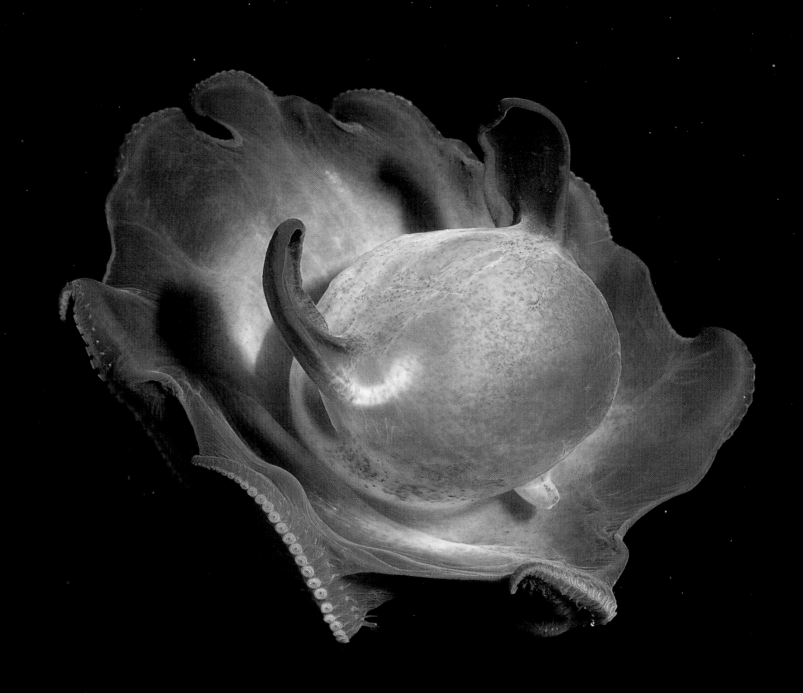

5 kg: Average weight of organisms per square meter near the surface. By comparison, the biomass at great depths is less than 1 g/m²; there, the populations are less dense, although the diversity of species is greater.

7 mm/year: Rate of expansion between tectonic plates under the Arctic Ocean. Compare this to the rate in the Pacific, where they separate at a speed of 18 cm/year—around twenty-five times as fast.

13°C: Temperature at the deepest point in the Mediterranean Sea, about 5000 m down. There, the abyssal waters are at least 10° warmer than in other oceans because the Strait of Gibraltar prevents cold oceanic currents from penetrating. The Red Sea holds the world's heat record at depth: 21.5°C at 3000 m depth.

60 m: Width of funnel-shaped deep trawling nets, larger than a soccer field.

20 km³: Volume of oceanic crust created each year where the tectonic plates separate.

150 million years: Time required for the Atlantic Ocean to attain its current size, starting from a crack in the continent.

300 to 500 times: Multiplier of nutrients in waters around hydrothermal chimneys compared to the rest of oceanic waters.

307 million km²: Area covered by oceans that are greater than 400 m in depth, which represents over 60% of the Earth's surface.

THE DEEP SEA IN FIGURES

9 m in 18 months: Growth record for a hydrothermal chimney, a black smoker nicknamed "Godzilla" by scientists from the University of Washington because of the impressive height it reached, nearly 50 m, before collapsing and beginning, immediately, to rebuild itself.

10 million years: Time it takes after a mass extinction for flora and fauna to return to their previous level of biodiversity.

75% to 95%: Percentage of animal species that expired during great mass extinctions. There have been five such events in the last 500 million years, with the most destructive instance occurring 250 million years ago.

100: Hydrothermal sites discovered in the last 25 years.

100 to 200: Shipwrecks found in waters deeper than 200 m. According to Robert D. Ballard, "the deep sea is the world's largest museum." It could easily contain several thousands, if not millions, of wrecks.

105°C: Optimum growth temperature for *Pyrolobus fumarii*, the world's most heat-resistant microorganism.

1000 years: Time required for water masses to circulate around the planet.

3729 m: Average depth of the oceans.

4188 m: Average depth of the Pacific, the deepest and largest of all oceans. By itself, it represents nearly half of the expanse of water on Earth.

1,500,000: Great whales killed by humans in the course of 200 years. Lucien Laubier calculates that the biomass of these whales is equal to that of 1.5 billion humans.

PRECEDING DOUBLE PAGE SPREAD
Grimpoteuthis sp.
Umbrella octopus
_SIZE 20 cm
DEPTH 300–5000 m

A startling vision of crimson, a deep-sea umbrella octopus hovers above the seafloor, descending occasionally to feed on the bottom or to attach its eggs to the branches of a cold water coral bush. By flapping its ear-like fins, the animal propels itself along at what seems a comically slow rate. In the abyssal plains, however, where food sources are meager, energy can never be wasted, and living life at a snail's pace is a common and successful strategy.

OPPOSITE
Amphitretus pelagicus
Telescope octopus
_SIZE 30 cm
DEPTH 100–2000 m

This remarkable octopus, looking a bit like a ghost wrapped in a transparent sheet of gelatin, lives exclusively in the midwater. It probably uses its rotating telescopic eyes to make out the silhouettes of prey situated above it. The biology of this unusual animal remains mostly unknown.

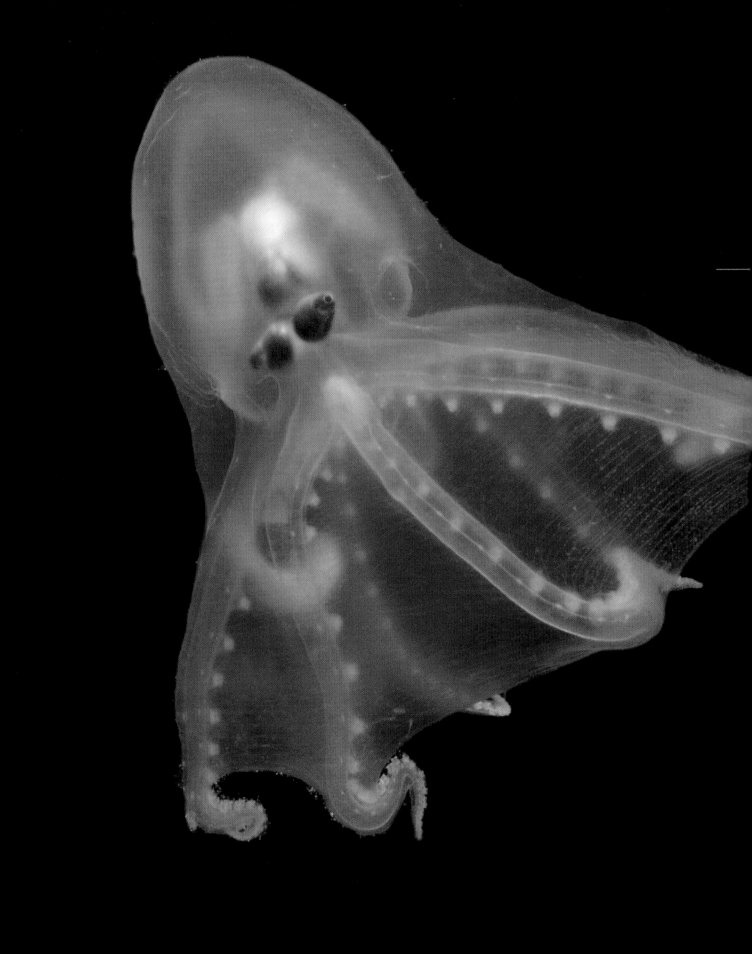

glossary

Abyss: Strictly speaking, the abyss is a particular zone extending between 3000 and 6000 m depth. By extension, the term is also used to designate the deep oceans overall.

Benthic: Pertaining to benthos (from the Greek, meaning "depth"), which designates the oceanic substratum. Refers to the fauna living on the bottom, in contrast to the pelagic fauna, which lives in the open ocean.

Biomass: The total quantity or mass of living material within a specified area at a given time. The concept of biomass allows one to express the idea of the abundance of animal presence in volume without having to use headcounts as is the case when speaking of density. Useful in that living organisms vary too broadly in size for density to be a meaningful measure.

Foraminifer: A living organism, often microscopic, surrounded by a calcareous or siliceous envelope. Foraminifers are abundant in sedimentary layers and provide information on the origin and evolution of the deep substratum.

Gonad: A reproductive gland.

Intertidal: Pertaining to the shore areas that alternate between submergence and nonsubmergence due to tidal oscillation.

Nutriments: Elements such as nitrogen and phosphorus, which are indispensable for plant growth and planktonic production. The oceans' fertilizers.

Oligotrophic: Pertaining to a nutrient-poor body of water, such as the central zones of the oceans which have very little plankton. From the Greek *oligo*, meaning "few," and *trophê*, meaning "nourishment."

Pelagic: Pertaining to fish and animals that live in the open sea, away from the sea bottom (from the Greek *pelagos*, meaning "open sea").

Photic: From the Greek *photos*, meaning "light." The photic zone is the space within lakes or oceans that is penetrated by sunlight. The lower limit of this zone depends on the particles in suspension in the water. In the open sea, it typically extends down to 200 m.

Phylum: One of the largest divisions of the animal or plant kingdoms. Within the hierarchy of taxonomic classification, a phylum is situated between kingdom and class.

Phytoplankton: Planktonic organisms belonging to the plant kingdom. Many of them are microscopic algae and diatoms (unicellular organisms) that photosynthesize, producing the first level in the oceans' food chain.

Precambrian: The very long geological period extending from the Earth's formation (some 4.5 billion years ago) to the Cambrian Period (about 540 million years ago), marked by the appearance in abundance of life, now recorded in fossils.

Upwelling: Upwelling occurs when surface waters sink to the bottom, slowly migrate across the deep oceanic basins, and rise to the surface. These waters contain nutrients from the sunken carcasses of small organisms. This surfacing thus brings nutrients to the surface, nourishing plankton, and is an important aspect of the marine food web.

Zooplankton: Planktonic organisms belonging to the animal kingdom. The majority are small crustaceans (copepods, krill), arrowworms, and gelatinous creatures that feed primarily on phytoplankton.

ABOVE
Harriotta haeckeli
Smallspine spookfish
_SIZE 65 cm
DEPTH 1400–2600 m

It's hard not to see elephant ears when looking at the spookfish's two immense fins. By making them undulate very slowly, the fish manages to remain perfectly stationary just a few meters above the bottom. The meaning of this midwater Zen exercise is yet to be deciphered. The spookfish spends the majority of its time around the continental slopes, where it uses its long flexible nose to search for prey in the mud.

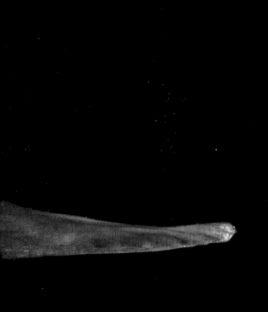

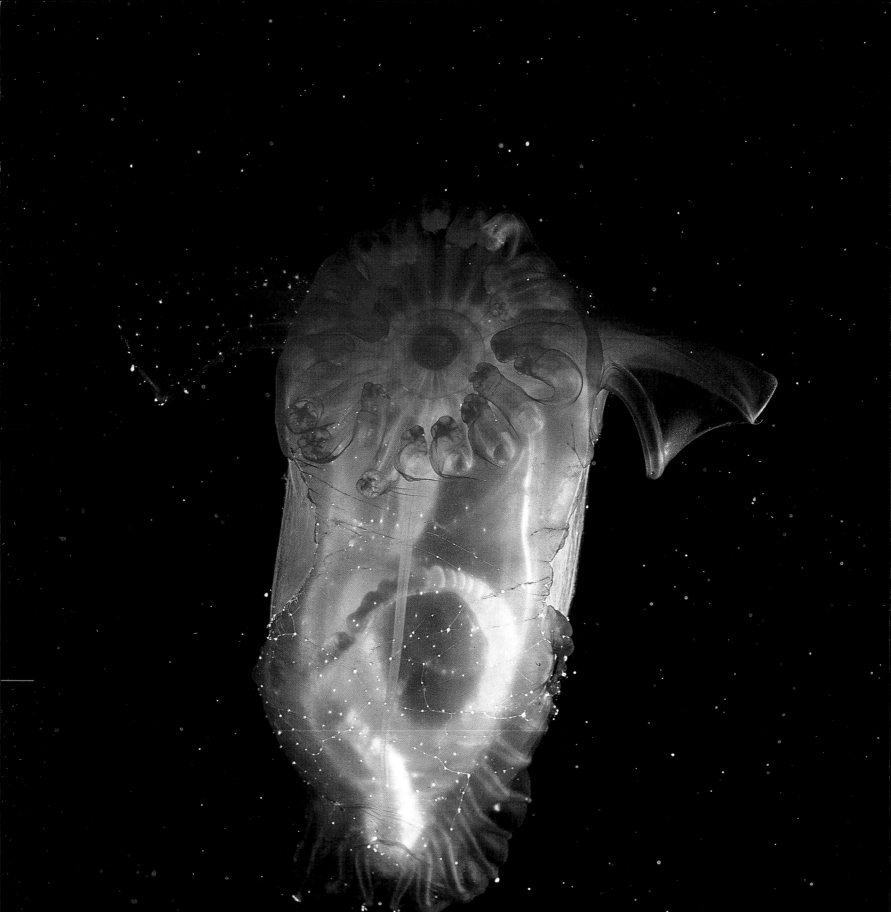

OPPOSITE
Enypniastes eximia
Deep-sea Spanish dancer
_SIZE up to 35 cm
DEPTH 500–5000 m

This deep holothurian is one of
a group of swimming sea
cucumbers; it undulates slowly
and gracefully through the water,
sometimes even rather far from
the bottom. The translucent
texture of the animal allows its
internal organs to be seen
from the outside; these consist
primarily of the digestive tube,
which filters the sediments
ingested, retaining the nutritive
particles.

PAGE 253
Grimpoteuthis sp.
Dumbo octopus
_SIZE up to 1.5 m
DEPTH 300–5000 m

The behavior and biology of
this finned octopus are still
largely uncertain. They are
frequently found close to the
bottom in all the world's oceans,
though they can also adventure
rather far within the water
column. The largest specimens
can attain 1.5 m in size.

Les Editions de l'Imprimerie
Nationale de Monaco, 1951
(first edition, 1902).

Ballard, Robert D.
*The Eternal Darkness: A Personal
History of Deep-Sea Exploration.*
Princeton University Press, 2000.

Batson, Peter.
Deep New Zealand. Canterbury
University Press, 2003.

Beebe, William.
Half Mile Down. Duell, Sloan,
and Pearce, 1951.

Byatt, Andrew, Alastair
Fothergill, and Martha Holmes.
*Blue Planet: A Natural History
of the Oceans.* DK, 2001.

Burnett, Nancy, and Brad Matsen.
The Shape of Life. Monterey Bay
Aquarium Press, 2002.

Hutchinson, Steven,
and Lawrence E. Hawkins.
Oceans: A Visual Guide. Firefly
Books, 2005.

Laubier, Lucien,
Des oasis au fond des mers.
Editions du Rocher, 1986.

Laubier, Lucien.
Vingt Mille Vies sous la Mer.
Editions Odile Jacob, 1992.

Macinnis, Joseph.
Aliens of the Deep. National
Geographic, 2004.

Matsen, Brad.
*Descent: The Heroic Discovery
of the Abyss.* Pantheon Books,
2005.

McKenzie, Michelle.
Jellyfish Inside Out. Monterey
Bay Aquarium Press, 2003.

Monod, Théodore.
*Bathyfolages: Plongées
profondes.* Editions René
Julliard, 1954.

Van Dover, Cindy Lee.
Deep Ocean Journeys. Perseus
Publishing, 1996.

Wrobel, David, and Claudia Mills.
*Pacific Coast Pelagic
Invertebrates.* Monterey Bay
Aquarium Press & Sea
Challengers, 1998.

Wu, Norbert, and Jim Mastro.
Under Antarctic Ice. University
of California Press, 2004.

Wyville Thomson, Charles.
*The Depths of the Sea: An
Account of the General Results
of the Dredging Cruises of H.M.SS.
"Porcupine" and "Lightning"
during the Summers of 1868,
1869, and 1870, Part I & Part II.*
Elibron Classics, 2003.

On my way to the Great Depths, a few individuals have been particularly helpful to me. They probably don't realize how important their spontaneous and friendly welcome of my ideas was for the realization of this project. First, I am naturally very grateful not only to all the researchers who agreed to write for the book, but also to Peter Batson, Lucien Laubier, Karen Osborn, Michel Ségonzac, Michael Klages, and Andreï Suntsov, whose advice, discussions, and proofreading of the manuscript were of inestimable value. I would also like to thank Henri Trubert, Martine Bertéa, Kim Fulton-Bennett, Steven Haddock,

answered my questions, my calls, who played along with the game simply because they liked my project—their lack of preconceptions or prejudices was not only very much appreciated, but also most fruitful: James Childress, Daniel Desbruyères, Jeff Drazen, Casey Dunn, Charles Fisher, André Freiwald, Kantaro Fujioka, Les Gallagher, Kristina Gjerde, Gary Greene, Russ Hopcroft, James Hunt, Emma Jones, Kim Juniper, Nicola King, Tony Koslow, Lisa Levin, Alberto Lindner, Dhugal Lindsay, Alexander Low, David Luquet, Larry Madin, Jérôme Mallefet, George Matsumoto, Mike Matz, Ian McDonald, Claudia Mills, Mark Norman, David Pawson, Dieter Piepenburg, Kevin Raskoff, Kim Reisenbichler, Bertrand

digital pictures without sufficient definition for the format of the book. I regret not having been able to satisfy all those who so kindly sent me their material from the four corners of the world. I thank you all very sincerely for having participated in this project, and do I hope that your magnificent photos will find the place they deserve in a future work.

I have unbelievable luck. To be able to do something I love as my profession as well as to be encouraged by those close to me, especially by my partner, Christophe, who has demonstrated immense generosity and patience beyond the ordinary in the face of all the sacrifices this project has imposed. And to be surrounded by kind and well-meaning souls who support me, actively or tacitly, through their constructive attention. All too seldom do I have the occasion to thank my grandmother, my mother and Denis, my father and Marie, Mathilde, Clothilde and Arnaud Nouvian, Denis Despretz, and my friends, especially Caroline Dickens, Mathias Chivot, Perrine Auclair, Céline Pissoort, Djamel Agaoua, Sabine Van Vlaenderen, and Benjamin Badinter.

Last but not least, my beloved, incredible, terrific sister, my "sœur terrible" Valérie, who always—in spite of the frantic rhythm of her life—finds the time to listen to me, to read my work, to tell me off, to make me laugh, to comfort me: her unshakeable confidence in me is the primary root of my energy. I'd also like to thank my painter friend Claire Basler, whose immense talent and freshness

have often revitalized me through the course of this project. When I surface up from my deep dives, my gaze falls on her superb oil paintings of forests hanging in front of my desk. Trees and seas: the perfect balance.

Thanks to Anne-Marie Bourgeois, who has passionately put heart and soul into this subject, creating a graphic setting matching the beauty of these creatures.

Thanks to David Batson for his incomparable effort in producing the very beautiful computer-generated illustrations.

What I will remember, long after the paper of this book has yellowed and the deep sea has become so accessible that this volume will seem obsolete, is the openness with which my project was received by researchers throughout the world. Without them, beyond any possible doubt, this book never would have come into existence; there's no mistaking that it is, above all, the fruit of their work. It is because of the trust that each one of them granted me that this project was really able to move forward. In my heart I will always cherish the great humanity of that confidence.

ACKNOWLEDGMENTS

Danièle Lemercier, Laure Fournier, Annelise Signoret, Joël Halioua, Nathalie Moritz, Guillaume Waelkens, Carole Caufmann, Christophe Hébert, and Claire Forest for their great support and cooperation.

I especially want to reserve a particular place for a very extraordinary man, Dr. Marsh Youngbluth of the Harbor Branch Oceanographic Institution. Thanks to him, I was able to dive to a depth of 1000 m in the *Johnson Sea Link* submersible. This experience was by far—by so very far—the most astounding moment of my life. The "thank you" I have for Marsh is well beyond words.

Thanks also to all who have given me their time, who have

Richer de Forges, Clyde Roper, Brad Seibel, Joseph Schrevel, Mark Schrope, David Shale, Rob Sherlock, Craig Smith, Erling Svensen, Rudolf Svensen, Tina Treude, Verena Tunnicliffe, Cindy van Dover, Michael Vecchione, Robert Vrijenhoek, Les Watling, Edith Widder, Craig Young, Richard Young, Naoko Zama, and all those I have not been able to mention here, but whose many contributions have been truly precious to me.

I gathered well over five thousand photographs for this book, but eventually had to select fewer than two hundred. Obviously such a choice was immensely difficult, and I had to sacrifice some superb images for lack of space or for technical constraints, such as various

A huge thanks to the Monterey Bay Aquarium Research Institute (MBARI), the Harbor Branch Oceanographic Institution, the National Oceanic & Atmospheric Administration (NOAA), and the Japan Agency for Marine-Earth Science and Technology (JAMSTEC) for letting me use their images in this work.

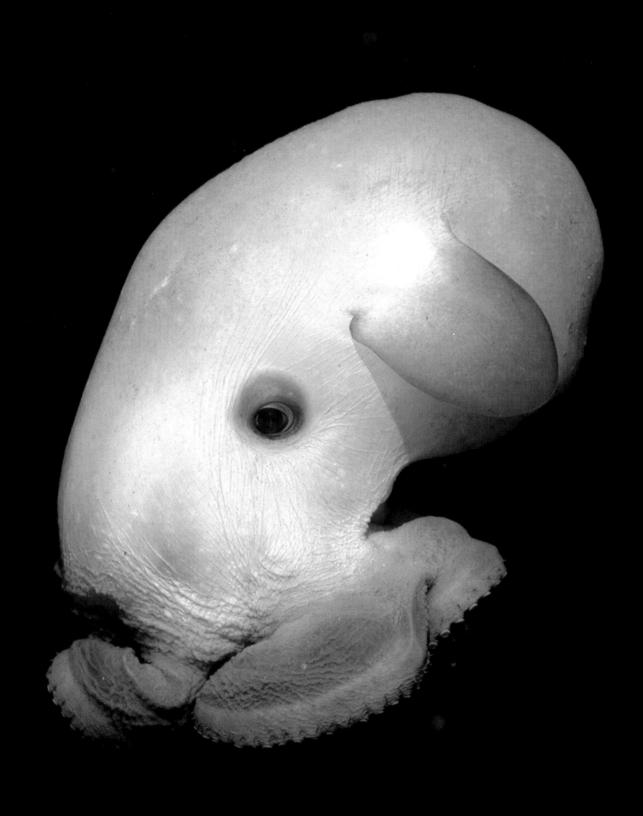

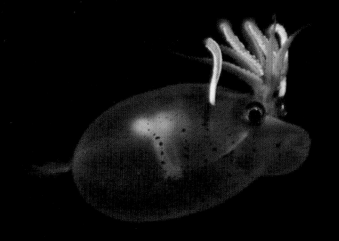

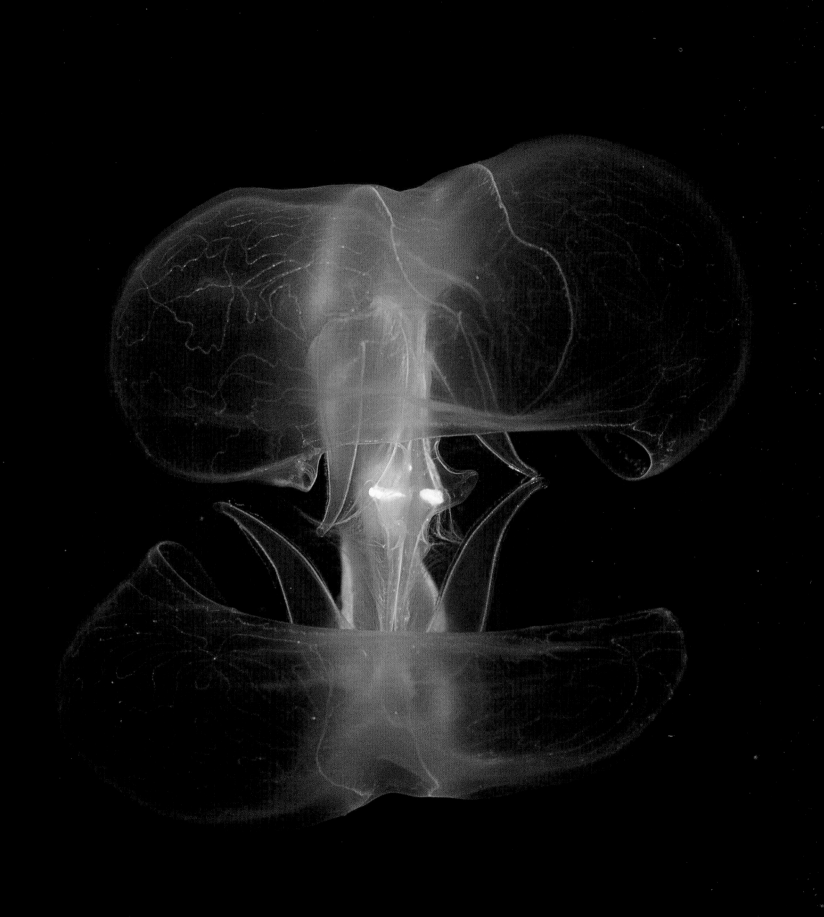

3D COMPUTER-GENERATED IMAGES

David BATSON
© David Batson, ExploreTheAbyss.com
123, 124–125, 184–185, 214–215

PHOTOS

Peter BATSON
© Peter Batson, ExploreTheAbyss.com
21 (2nd from the left), 21 (3rd from the left), 54–55,
119, 223 (center left), 242–243, 244, 245

Derk BERGQUIST
Derk Bergquist © Penn State University
227, 231 (bottom right)

Jeffrey DRAZEN
© 2005 Jeffrey Drazen
38

Charles FISHER
Charles Fisher © Penn State University
222 (top), 223 (top left), 223 (center right),
230 (top), 230 (center), 231 (top left),
231 (top right)

Per FLOOD
Per R. Flood © Bathybiologica A/S
75 (center), 113, 116, 139

Jan Helge FOSSAA
© Jan Helge Fossaa, IMR
22 (bottom)

Steven HADDOCK
Steven Haddock © 2006 MBARI
110–111, 155
Steven Haddock © 2005 MBARI
4, 5, 21 (4th from the left)
Steven Haddock © 2004 MBARI
36–37, 60, 61
Steven Haddock © 2003 MBARI
30–31, 75 (top), 88, 117 (left), 247
© Steven Haddock © 2002 MBARI
108–109
© Steven Haddock 2001
67, 107

CREDITS

© Steven Haddock 2000
44
© Steven Haddock 1998
80–81

Harbor Branch Oceanographic Institution
© 2006 Harbor Branch Oceanographic Institution
24 (4th from the top)

Russ HOPCROFT
© Russ Hopcroft / UAF
27

Hawaii Undersea Research Laboratory
Courtesy of the Hawaii Undersea Research Lab.
Submersible pilots: Terry Kerby and Colin
Wollerman. Principal investigators: Hubert Staudigel
& Craig Young
208–209

IFREMER
© Ifremer
24 (3rd from top)
© Ifremer/Campagne Biozaïre 2, 2001
165 (center right)
© Ifremer/Campagne Phare 2002
220
© Ifremer/A. Fifis
221
© Ifremer/Campagne Hope 1999
222 (center)
© Ifremer/Campagne Atos 2001
223 (top right)
© Ifremer/Campagne Exomar 2005
223 (bottom right)

Image Quest 3D
© Peter Herring/imagequestmarine.com
21 (1st from the left)
© Peter Herring/imagequestmarine.com
101
© Roger Steene/imagequestmarine.com
156–157

JAMSTEC
Kaiko © JAMSTEC
24 (5th from the top)
© Dhugal John Lindsay, Ph.D.(JAMSTEC)
74

Miriam KASTNER
© Miriam Kastner
233 (bottom)

Lisa LEVIN
© Lisa A. Levin
152

The Stephen Low Company
© The Stephen Low Company
25, 210, 216–217, 218–219

Lawrence MADIN
© L. P. Madin, WHOI
159, 250

Marine Themes
© marinethemes.com/Kelvin Aitken
8–9, 190, 191, 192–193

George MATSUMOTO
G. I. Matsumoto © 1988 MBARI
28–29
G. I. Matsumoto © 2003 MBARI
32–33, 42–43
G. I. Matsumoto © 2002 MBARI
114
© G. I. Matsumoto 1985
170, 255

Mikhaïl V. MATZ
© Mikhaïl V. Matz
75 (bottom)

Monterey Bay Aquarium Research Institute (MBARI)
© 1999 MBARI
207
© 2000 MBARI
24 (6th from the top), 201 (center left),
© 2002 MBARI
203, 126, 200 (bottom), 196, 201 (top right),
© 2003 MBARI
187, 254, 70, 201 (center right), 201 (bottom left),
253, 26, 231 (center left),
© 2005 MBARI
164 (top), 13
© 2002 MBARI/NOAA
201 (top left), 200 (top), 23,

Ian McDONALD
© Ian McDonald
165 (center left), 165 (top right), 228 (top), 228
(bottom), 230 (bottom), 231 (bottom left), 231
(center right), 233 (top)

Claudia MILLS
© Claudia Mills
188–189

National Geographic
© William Beebe/National Geographic Image
Collection
24 (top)
© Bill Curtsinger/National Geographic Image
Collection
169

Natural Visions
© Peter David / Natural Visions
132, 138 (center)

NOAA (National Oceanic & Atmospheric
Administration)
Image courtesy of the Hidden Ocean Science Team,
CoML & NOAA
18

Karen OSBORN
K.J. Osborn © 2005 MBARI
59
K.J. Osborn © 2004 MBARI
91

Courtney PLATT
© Courtney Platt
150–151

Kevin RASKOFF
© Kevin Raskoff
66, 69, 90, 117 (right), 134, 135, 166, 172,
177 (top), 177 (bottom)

Kim REISENBICHLER
Kim Reisenbichler © 1996 MBARI
129

Research Center for Ocean Margins
© Research Center for Ocean Margins
224

David SHALE
© David Shale
48, 52–53, 73, 92–93, 120, 136–137, 248
© David Shale / Claire Nouvian
hard cover, 14–15, 46–47, 58, 64–65, 83,
94, 95, 96–97, 191

Rob SHERLOCK
Rob Sherlock © 1998 MBARI
19

Craig SMITH
© Craig R. Smith
234, 235, 237

Verena TUNNICLIFFE
© Verena Tunnicliffe & Kim Juniper
41
© NOAA Ocean Exploration & Verena Tunnicliffe
165 (bottom right), 171 (bottom),
213, 222 (bottom), 223 (bottom left)

University of Aberdeen
© Oceanlab, University of Aberdeen
238 (top), 238 (center), 238 (bottom), 239

UW Photo
© Nils Aukan / uwphoto.no
146–147
© Bjørn Gulliksen / uwphoto.no
16–17, 174–175, 179
© Erling Svensen / uwphoto.no
144–145, 173, 176, 199, 204–205
© Rudolf Svensen / uwphoto.no
206

Wim VAN EGMOND
© Wim Van Egmond
106, 130–131

Les WATLING
© Les Watling for the Mountains in the Sea Research
Team, IFE, URI-IAO, and NOAA
22 (top), 164 (center), 164 (bottom), 165 (top left),
165 (bottom left), 171 (top), 171 (center), 200
(center), 201 (bottom right)

Edith WIDDER
© Edith Widder
21 (last from the left), 51, 87, 102, 105, 138 (top),
138 (bottom), 140–141

David WROBEL
© David Wrobel
10, 20, 34–35, 56–57, 76–77, 78, 79, 98–99, 112,
158, 160–161, 180, 183, 194–195

Craig YOUNG
© Craig M. Young
142–143, 148, 162, 163

Marsh YOUNGBLUTH
© Marsh Youngbluth, Harbor Branch Oceanographic
Institution
62–63, 84, 149, 186

In the Public Domain
U.S. Naval Historical Center Online library
24 (2nd from top)

PAGES 4–5
Unidentified species
_SIZE 3–25 cm
DEPTH 1200–1800 m

Resembling two UFOs floating
in space, these delicate comb jellies
have yet to be described and named
by biologists. Gelatinous animals
of the deep are often larger than
their alter egos that live in the sunlit
layers of the open ocean. The cold
and dark abyss is a peaceful haven
for fauna repelled by the tumultuous
surface waves.

PAGES 8–9
Chlamydoselachus anguineus
Frilled shark
_SIZE 2 m
DEPTH 1600 m

One of the most primitive living
sharks, the frilled shark is the sole
surviving member of a family
otherwise known only from fossils.
Its three hundred impressive
three-pronged teeth are not used for
eating hard-shelled prey, but rather
for soft creatures, mainly squid.

PRECEDING DOUBLE PAGE SPREAD, LEFT
Helicocranchia sp.
Piglet squid
_SIZE up to 10 cm
DEPTH 400–1000 m

This powder-puff squid, with its
globular body and its small caudal
fin, certainly does not appear
to be an animal built for speed.
The larvae of this animal develop
in the food-rich waters of the surface
and, guided by instinct, start a
downward migration as they mature
until they reach the depths, where
they live as adults. Once an adult,
the squid remains at depth and no
longer migrates toward the milder
waters of the photic zone.

PRECEDING DOUBLE PAGE SPREAD, RIGHT
Bolinopsis sp.
_SIZE 20 cm
DEPTH unknown

Discovered in Antarctica, this
completely transparent ctenophore
swims in a vertical position, keeping
its two lobes open like a toothless
jaw. It captures other ctenophores
as well as copepods and planktonic
shrimps. This new species is still
awaiting description.

CHIEF EDITOR Henri Trubert   ADVISORY EDITOR Peter Batson   COORDINATOR Elise Roy   INTERNATIONAL RIGHTS Martine Bertéa
WITH THE ASSISTANCE OF Mathilde Demarcy AND Catherine Farin   PHOTOGRAPHIC RESEARCH Claire Nouvian/Association Bloom
TRANSLATIONS FROM THE FRENCH BY John Venerella   GRAPHIC DESIGN Anne-Marie Bourgeois/m87design   LAYOUT A.-M. Bourgeois
WITH THE ASSISTANCE OF Ilanit Wouz   PRODUCTION Frédéric Sauzé   PHOTOENGRAVING APS/Chromostyle, Tours (France)